Interdisciplinary Design

New Lessons from Architecture and Engineering

Edited by Hanif Kara and Andreas Georgoulias

Contents

Foreword
Jorge Silvetti   Effective Affinities   4

Introduction
Hanif Kara   Redesigning Attitudes   10
Andreas Georgoulias   Learning from Design   16

Chapter 1. Structure and Design
Sabrina Leon   Structure as Architectural Intervention   40
Burak Pekoglu   Behind the Curtains: Backstage   55
Murat Mutlu   Surfing the Wave   59
Evangelos Kotsioris   Toward a New Sobriety: Rebel Engineering with a Cause   65

FOA / AKT   Highcross Leicester   70
FOA / AKT   Ravensbourne College   74

Francisco Izquierdo,
Giorgi Khamaladze,
Jarrad Morgan,
Stephanie Morrison   GSD Lite   78

Travis Bost,
Werner Van Vuuren   Harvard M.O.D.E.   86

Zaha Hadid Architects /AKT   Phaeno Science Center   92
Christos Passas   Design the Cloud: Cross-disciplinary Approach to Design   94

Chapter 2. Contextual and Systemic Design
Dimitris Papanikolaou   Changing Forms, Changing Processes   106
Sylvia Feng   Enhancing Prefabrication   115
Scott Silverstein   Design for Disassembly: Closing the Materials Loop
                                            without Sacrificing Form   119
Azadeh Omidfar,
Dan Weissman   Design with Climate: The Role of Digital Tools
                                            in Computational Analysis of Site-Specific Architecture   125

Foster+Partners / AKT   Masdar Institute of Science and Technology   132
Chris C. L. Wan   The Role of the Client in the Design of Sustainable Developments   134

Cara Liberatore,
Lesley Mctague,
Anthony Sullivan   Open House   144

Charles Harris,
Ji Seok Park   Refabrication   150

Feilden Clegg
Bradley Studios / AKT   Heelis - National Trust Headquarters   156

Guy Nevill   On Heelis: The Role Played by Environmental Engineers   158

Chapter 3. Computation and Design Optimization

Lee-Su Huang	Algorithms in Design: Uses, Limitations, and Development	166
Jessica Sundberg Zofchak	Integrated Design: A Computational Approach to the Structural and Architectural Design of Diagrid Structures	173
Fernando Pereira Mosqueira	Form Finding: The Engineer's Approach	182
IJP Corporation / AKT	Henderson Waves Bridge	188
George L. Legendre	Henderson Waves: A Collaboration	192
YueYue Wang, Hailong Wu, Zhu Wu	Double Shell	200
Vera Baranova, Sophia Chang, Bernard Peng	Jewel Box	206
Heatherwick Studio / AKT	UK Pavilion, Shanghai Expo 2010	212
Katerina Dionysopoulou	In collaboration…	214

Chapter 4. The Emergence of a New Discipline

Jennifer Bonner	Death of The Star Architect	226
Svetlana Potapova	How the Architect Found the Engineer	231
Fai Au	A Necessary Resistance within Architect–Engineer Collaboration	236
Advait M. Sambhare	Interdisciplinary Collaboration: Enabling Architects to Regain Leadership in the Building Industry	243
AHMM / AKT	Adelaide Wharf	250
Morag Tait	Collaboration as a Working Process	254
Andrew Pedron, Nathan Shobe, Trisitie Tajima	Center for Advanced Architecture	258
Danxi Zou, Song He, Bo Feng	Bridge Gallery	264
Feilden Clegg Bradley Studios Ian Taylor	Westfield Student Village	272
	Climate Comfort Collaboration	275
	Contributors	278
	Acknowledgements	284
	Image credits	286

Foreword:
Effective Affinities

Jorge Silvetti

It is in the nature of architecture to be perennially interested in establishing active inter-actions with other disciplines of knowledge and art. For those engaged with design—as a goal-oriented, creative formal practice grounded in the interpretation of incomplete, biased, and sometimes even contradictory information about myriad real-world vari-ables, reliant on intuition and fantasy as much as on fact—it is "natural" to yearn at times for reinforcement from other disciplines whose objects of interest overlap with some aspects of architecture's wide-ranging concerns. Various intellectual domains and their corresponding technologies have acted at one time or another as strong magnets for architecture (and as often as objects of veiled envy). Indeed, throughout its history architecture can be found in friendly relationships with other practices, and a history of how different periods have defined "what architecture is" would certainly yield an extraordinary parade of distinct instances not so much about its products—which remain stubbornly of the same kind—but about how it has articulated its alliances to establish its historical position in relation to the defining issues of an epoch. But that is a topic for another occasion.

Where I'd rather focus instead, on this occasion of introducing the work of Hanif Kara, Andreas Georgoulias, their professional collaborators, and their students at MIT and the Harvard Graduate School of Design, is on what we have learned from architecture's history of alliances and their corresponding polemics—represented best by the oscilla-tion between overenthusiastic engagements and obdurate isolationisms—and how that history can help us to correctly place their work and understand the significance of their contribution.

And to begin, let's underline this stressful, general condition that characterizes the environment within which these interactions usually occur: the vacillating swings be-tween eagerness and restraint, enthusiasm and fear, conviction and doubt about what these liaisons promise or how they might threaten architecture. This volatile, unresolved state of mind no doubt has many roots, but we can expose just one that serves well to illustrate, factually and metaphorically, the confusion between the nature of the prod-ucts of architecture and the processes through which we arrive at them. Because if the former has been and continues to be (and better be!) large, complex physical entities of a stable nature—heavy and durable and consequently highly conspicuous, relentlessly present as points of reference in daily life—the latter, architecture's disciplinary position vis-à-vis other domains of knowledge and practice, is a rather unstable condition, in con-stant flux, with a historical variability that has seen it associated at different times with mathematics, music, geometry, natural or social sciences, operations research and cy-bernetics, ecology, linguistics, and history, to name just a few. The inevitable perplexing question is: How can the long-lasting, unmovable objects of architecture be based on a practice that seems in constant instability as to its intellectual identity?

Perhaps it is this tension between these two spheres of experience, between the stability of the products we build—which exist on their own, for all to share—and the perceived fickle condition of our disciplinary status among the constellation of practices that define academia and the profession, which ultimately fuels old-age battles between those who demand autonomy and others who insist on integration.[1]

Yet those two spheres—the one of daily life experience and the other of intellectual disciplinary practice—are neither analogous nor interchangeable at the moment we try to define their content, the former being a physical and symbolic mixture of cultural determinants, the latter a methodological and discursive practice in pursuit of knowledge and technologies with corresponding symbolic languages and culture. We've known for awhile, at least since the end of the period when the discourse on architecture's autonomy was highly focused, in the 1980s,[2] that this is a false dichotomy—that while at any given time architecture must insist on a disciplinary territory of its own, at that very same moment and precisely in order to stake out such a position, it must also define and articulate its historically determined, contingent, and necessary ties, dependences, and dominances with other realms of knowledge and reality. What in fact should be at stake in a polemic such as this is not whether architecture should exist in a pure self-reflective state that inoculates it against external infections (impossible) or whether it should instead lower its defenses and welcome the beneficial contagion of other agents (inevitable): rather, it should be about how we ought to best articulate those moments of re-accommodation to define both new domains of autonomy and the neighboring territories of knowledge that would constitute the productive disciplinary environment within which we would operate and interact.

In the time that I have been involved with architecture, I can recall at least three major episodes dominated by intense interactions between architecture and other disciplines, when calls for intellectual and methodological cross-fertilization were heard, experimentation conducted, and great expectations developed. Two of these engagements, arising while I was a graduate student at the University of California at Berkeley, came from different academic streams, but for awhile they partially overlapped in time and content, stimulated without doubt on the West Coast by the encouragingly multidisciplinary atmosphere of the recently created "College of Environmental Design" on the UC campus.[3] The first (clearly American/Anglo-Saxon) strain, enthralled by the academically popular "scientific method," was pushing for an integration of all "environmental disciplines" with the social sciences, following the models of general systems theory and cybernetics, while betting eagerly on the promise offered by the incipient but already influential computational techniques that were making their first entries into architecture.[4] The other, a distinctly European strain, aimed equally at bringing the social realm within architecture (but clearly avoiding the "social sciences" label), in this instance not through the "scientific method" but rather through an activation and actualization of the concept of culture, following primarily contemporary developments in linguistics and anthropology.[5] The intensity and rigor with which these strains worked their ways into and through architectural academia, the professional press, and ultimately practice itself are unimaginable today, but can perhaps be best measured by the success of their most conspicuous and lasting product: the totally reconfigured—indeed the invented—field of architectural theory as a nonprescriptive, critical, independent discourse that remains the engine of many academic and research endeavors.

Inevitably, that intensity and rigor also revealed the pitfalls that interdisciplinary efforts in architecture have generally encountered whenever they become dominant and their strategic rules of engagement are not well understood. In particular, confusion results when models and methodologies are transferred without transformation from one field to another, a process that renders architecture a mere representational system of something else and generates highly metaphorical discourses. Thus in the recent past architecture has not only been analogized to linguistics or nature—to name just two preeminent cases—but has been taken beyond the analogy and effectively understood by wide sectors of academia and the profession as operating like a language or a natural organism. Then the conceptual apparatus, which had served so well at the beginning to amplify the understanding of architecture, collapsed in the midst of a labyrinth of mirrors, and it became difficult to distinguish between what was architecture and what was represented. The causes of these distorting, naïve, negative outcomes are many, but they can all be traced to architecture's inability at any of those occasions of disciplinary alliance to figure out how to delimit its proprietary territory as a multidimensional intellectual and pragmatic construct that operates simultaneously at multiple levels of action, inquiry, and imagination and with different degrees of porosity to external factors—but with a clear territorial definition. It always seems difficult for architecture to control its enthusiasm and monitor the limits of the metaphors it is so prone to depend on for its discursive practices.

The lull that followed this period (basically the last decade of the twentieth century, when architecture temporarily abandoned the strategy of interdisciplinary engagement) was soon to be shaken by the impact of digital and information technologies' entry onto the scene. Among the many altered conditions that it created, the most remarkable was the sudden and seemingly miraculous reversal of the process of design, as we were granted the novel ability to generate complex and formerly unimaginable forms a priori, without the "crutches" that goals, programs, or objectives provide, leading to a new kind of enthusiasm and optimism (albeit accompanied by the anxiety of exhibiting formal dexterity without any reasonable or convincing content). And while the reconfiguration of the field of possibilities of architecture is still going on, these startling new opportunities have already reopened the doors to free associations induced by the sudden and surprising demise of some of the simplest formal constraints of architecture, ingrained and accepted since time immemorial. Possibilities for formal and technical resolution that had been deemed unattainable, such as the ability to smooth sharp edges and to distort the grid in any direction and throughout any surface, came at no extra cost in time or money. A corollary to many of these new options appeared: the apparently oxymoronic condition of having modules that are not dimensionally constant but able to be repeated following topological but not geometric logic.

These startling new conditions of formal possibility—presented to us not as the culmination of a traditional logical process of design but as the independent outcome of a technical ability that allowed the production of unimagined formal portents a priori and for the asking—generated an exhilarating mixed sense of freedom and apprehension, as architecture had to reckon with its inability to overcome the paralyzing condition of producing complex forms without content. An initial innocent albeit exhibitionist period dominated by the attitude of "we make it because we can" produced nameless, cheerful, innocuous blobs.[6] But these soon gave way to a sense of urgency that reopened the doors of architecture to other disciplines more able to provide either the technical and intellectual assistance to channel these anxieties properly or the rhetorical devices to interpret these novel forms as signs of a new definition of architecture.

Not surprisingly, the latter quickly took the lead along the well-trodden path that academia is most prepared for; once again metaphorical thinking took over, as if the smoothness of the formal gradients that we had created would ease the slippage from one intellectual domain to another. The "illustrative" power of architectural form that had dominated so much of the postmodern era (including the "deconstructivist" moment that we thought was its concluding episode) again reared its ugly head and gave us a trivialization of the idea of interdisciplinary work in the form of forced weak associations. The scenario was propitious, if not inevitable: the future world was promised to exist in an environment where all was smooth, continuous, undulating, and undifferentiated, where forms, domains, genres, infrastructure, public spaces, and buildings all blended and folded into one another. It was tempting to think that there would be no more specificity to the design disciplines or, to speak metaphorically, there would be no boundaries between them, as everything could slip into everything else, resulting in a continuous territory that became either the realm of an infinitely undefined hopeless landscape or, more likely, just the deceiving endlessness of a world of reflections in a hall of mirrors. Undoubtedly, the pendulum had swung back to the extreme of overenthusiastic engagement as the third episode of interdisciplinary interaction got under way.

In fact, and because of the peculiar ability we acquired to make fast connections and slippages, to operate with smooth navigational processes and to gain easy access to just about everything, this particular direction did not engage any sensible and rational discussion about the pertinence that a new lineup of potential allied disciplines might bring to architecture's table. Rather, this was an entirely new game, vague but intense as the mood of the times would have it, in which all disciplines suddenly seemed equally important and relevant to us, thus making the playing field rather bland, the selection random, and discussion not necessary. This total availability, and hence the low intensity of demand, focused principally on the "connections" themselves rather than on the content. Eventually the quest for interdisciplinarity became the goal in itself, its structure and its content—thus the adoption of quickly concocted terms to describe new disciplinary realms (real or imaginary) with which academia has bombarded us lately as "new" directions that urgently ought to be followed: infrastructural ecology, landscape urbanism, landscape planning, ecological urbanism, etc., which have taken the place in the lingo of academia of the once insufferable highbrow neologisms of cultural theory.[7] The jury is still out of course, and many good points have been made about the need to give consideration to some such marriages, but the meager results are at best a mixed bag of promising and trivial findings. What is indisputable though is that once again the central discussion about how we undertake the reconfiguration of our field of autonomy and our field of interactions has been mostly, and irresponsibly, lacking.

It is within this unsettled historical context that we must welcome Hanif Kara, Andreas Georgoulias, and their collaborating practitioners' initiatives in working with students from the GSD's architecture and MIT's engineering programs as an encouraging first sign that the panorama of interdisciplinarity may be getting less foggy and the view more focused. With this work, we are surely in the territory of opportunities opened up by the rekindling of the idea of interdisciplinary collaboration. Their sober, measured, and thoughtful approach to establishing a mature and mutually enriching relationship among old friends stands out in clear contrast to other ongoing attempts that at best repeat the oscillation between opposites and at worst escalate into claims of superiority with undertones of unfriendly takeovers. In this climate, it is refreshing to see that Hanif begins the book by reminding us that "in the struggle for reinvention, all disciplines have tended to 'peek' into each other's work during shrinking markets or major changes in the economic order. . . redrawing disciplinary boundaries as a means of survival," followed by the warning that these processes, when purely opportunistic, lead to confusion. He recommends instead that the need for successful interaction requires "each participant having both a deep knowledge of his or her own discipline and a receptivity to the equivalent knowledge of others." Such a requirement, in an academic context, I trust to mean that as we take on new partners we must consciously and purposely strengthen our own identity by coming prepared with our own riches.

If to some this latter dictum reads as too general and obvious a requirement for any disciplinary interaction, it becomes quite poignant when considered in the light of those current attempts mentioned above that negate identities and shy away from disciplinary rigor. What is key in the process that Hanif, Andreas, and their team chose as their basis of pedagogy is not only the need to counter the isolationism inherent in all purist disciplinarians' defense of the integrity and autonomy of their métiers as well as uncontrolled disciplinary spillovers, but more important, that they chose intelligently to understand how, de facto, "interdisciplinary activity" takes place in the real world. Thus the most significant move would be to bring their approach closer to the scenario that urgently requires the confluence of many an expertise, chiefly among them those of architects and engineers of all kinds.

Strategically, today, this is very sound—unlike two decades ago when a young and vibrant Theory of Architecture, from its enthroned position in the academic setting, led the way for all design experimentations and creativity while practice followed, panting, a few steps behind. Today's advances in material and digital technology, which must be acknowledged as the current engines of advancement, find their most progressive and productive locus of opportunities in the world of design and building practices. So, while superficially the history of rapturous attractions and hurried marriages appears to repeat itself, what we have in this case is the cultivation of mutual mature attractions followed by a smart courtship—indeed an entirely new framing in which interactions do not occur as result of speculative thinking but are modeled on what actually is taking place in the most advanced conditions of practice today—a model that Hanif can bring with confidence and command given his spectacular involvement on those frontiers of action. To put it in other words, the ingredients continue to be the same but their strategic, methodological, and philosophical deployment are what differentiate the process that is the basis of this book.

But I should now let the book, its authors, and its many illuminating cases speak for themselves. It is a compelling array of not just "research cases" but also of "stories," of unique moments where the tools of research and its protagonists, with the immediacy of their enthusiasms, hesitations, anxieties, and joys, are vividly and convincingly present in a more appropriate narrative method of reporting research in the field of design where uncertainty, intuition, and affects are as indispensable conditions as the adherence to strict methodological rigor. Indeed, this work may very well contain the seeds of the model of research in architecture that we have long been striving to uncover and explain to the rest of the academy, which has tended to be doubtful about our belonging to the world of scientific and scholarly conduct. It is precisely in this present uncertain world, where those old paradigms of academic propriety are questioning their own rectitude, that the fresh and inspired confidence of engineering—as it courts and engages architecture's idiosyncratic ways of knowing and producing in the forum of ideas and actions—may be the most optimistic way to map out new and promising paths of research. For once the seemingly incompatible modes of operation of architects and engineers, the former following leaps of logic and the latter the certainty of facts, could make a persuasive case for interdisciplinary interaction and put to rest the tired polemics of autonomy versus integration. And with their realms unthreatened, there should be no fear of a return to dry and uninspired relationships "by necessity" between design and technology: the mutual interests between architecture and engineering that have been activated in the work here presented are based on genuine attraction. They have found in the other not a complement but a common trait that is articulated in a different fashion: they are interested in not just using the other's exclusive knowledge but, more important, in what they have in common but interpret differently. Ultimately, let's not forget that history itself clearly favors this partnership. Not all interactions are created equal, and insofar as architecture is concerned, engineering has, by right, a sound claim for primacy in the reckoning of architecture's favorite partners.

Notes

1. Or other current terms for integration: interdisciplinarity, multidisciplinarity, transdisciplinarity, etc.
2. This polemic found its most productive and active locus in the United States at the Institute for Architecture and Urban Studies (IAUS) in New York, where the influential work and ideas of Robert Venturi, Aldo Rossi, and Peter Eisenman were mostly interpreted and discussed in this key.
3. The College of Environmental Design was created in 1959 and claims to be the first to combine the academic disciplines of architecture, landscape architecture, and city and regional planning and to "lead the way toward an integrated approach to analyzing, understanding, and designing our built environment" (from their current website). While other combinations of similar disciplines have been adopted in many places, the emphasis on this "new configuration" and the choice of a name with much broader reach were total novelties at the time and catapulted Berkeley and its promotion of multidisciplinarity during the late 1960s to the forefront of design education and research around the world.
4. The thinking and work of Herbert Simon and Karl Popper being dominant.
5. Among the leading minds corresponding to this strain we can cite Ferdinand de Saussure, Claude Lévi-Strauss, Roman Jakobson, and Roland Barthes.
6. For an extended discussion of this particular period in architecture's recent history, see Jorge Silvetti, "The Muses Are Not Amused—Pandemonium in the House of Architecture," in *Harvard Design Magazine*, no. 19 (2004), special issue on Architecture as Conceptual Art.
7. While some of those pairings have proven to be a welcome temporary mechanism to provoke discussion and clarification about the many disciplines involved, their proliferation and stridency—all arising within a short period of time—make them at best symptoms of confusion and at worst the result of intellectual turmoil rather than of patient and systematic academic argumentation.

Redesigning Attitudes

Hanif Kara

Architects and engineers both claim to be designers, though how they define design and the approaches they use to realize it vary widely. Their interaction, however, has created some of the world's most memorable, enduring, and impressive buildings. The explosion of digital technologies illuminates the complexity and nonlinearity of the process that these designers experience daily, while massively expanding the ability to visualize and represent forms and to analyze their structural engineering behavior. Technology has obviously changed both architecture and engineering, and so also the potential for interaction. That is the subject of the research and discourse presented here.

In a course at the Harvard Graduate School of Design attended by graduate students in architecture and MIT graduate students in structural engineering and computation, students and instructors examined a series of built projects over several years to cover many different design intentions, budgets, and purposes. What emerges is an appreciation of the breadth and depth of ways in which architecture and engineering can interact, rather than fulfilling the assumption that one might have held twenty or thirty years ago—that sophisticated engineering necessarily means complex forms, or that engineers cannot contribute to simple forms. Studying built projects where the instructor has personal knowledge can trace that interaction (project to project) and so provide a set of prisms for investigating the relationship between architecture and engineering—and the nature of design itself. The projects and methodology used in the discussions provide a shared context to provoke thinking about design, develop new viewpoints, and, through communication across disciplinary boundaries in teaching, practice, and construction, unlock the potential to discover new meanings.

Motivations and Intentions
In the struggle for reinvention, all disciplines have tended to "peek" into each other's work during shrinking markets or major changes in the economic order (e.g., the industrial revolution), redrawing disciplinary boundaries as a means of survival. The course we refer to started in 2006 as an experiment at a time when markets were expanding and architecture faced the challenge of responding to new popular imaginations, for little or no purpose at times. Many engineers have exploited this emblematic power of architecture to reinforce their own positions. In my view, this sometimes creates simple estimates of the work of architects in an atmosphere of unclear thought. If this misreading remains unanalyzed and unchallenged, it leads to a confused experience in practice.

This narrative deals with a limited period, between 1996 and 2010, and a particular viewpoint, so the projects are recent and represent stages in the ongoing evolution of their architects' oeuvres, as well as in the development of engineering techniques. It attempts to identify any confusion and correct it. For that reason, my long and deep immersion in the field of structural engineering is a more appropriate intellectual base for the discourse than a historical analysis of the relationship between engineering and architecture. Although the knowledge acquired through that immersion becomes manifest in specific examples of engineering design, it is broad and not skewed by a quasi-moral sense of what constitutes "good" engineering.

Successful interaction depends on each participant having both a deep knowledge of his or her own discipline and a receptivity to the equivalent knowledge of others. I call this interdisciplinary working because it encourages a better understanding of the nature of the interaction and an appreciation of when an understanding of one's own discipline ceases to be applicable. This works well in an environment conducive to research, whether in practice or within a university.

The potential for interdisciplinary working does not have to be realized, however, and moreover, it can be realized in many different ways. Rather than seeking to unravel the two disciplines and define them in abstract terms, one has to examine the syntheses they have created. Andrew Saint pinpointed why architects and engineers need to learn about each other: "There is a deeper reason why architects learn something about engineering. Structure is a basic requirement for any design that is to be built. Because of that it has in some sense to be confronted, incorporated and, quite possibly expressed. In those terms an architectural design that does not address structure is incomplete or illogical. One of the many tasks of an architectural education is to promote an informed attitude towards structure."[1] In my experience, engineering educators rarely think this way, while some architectural schools today also need to consider this. This research aims to reinforce such threads in the fabric of the relationship between the disciplines for the sole purpose of producing better buildings. The intention of the research is to refashion how this relationship can be seen, through projects that are themselves filtered through my experience of learning as a student, teaching, studying, practicing structural engineering, and for fifteen years serving as a principal of a successful independent design studio that specializes in structural engineering design.

Built Projects as Instruments
In "Politics and the English Language," George Orwell wrote: "The English language becomes ugly and inaccurate because our thoughts are foolish, but the slovenliness of our language makes it easier to have foolish thoughts."[2] In much the same way, this research tries with all humility to avoid "slovenliness" of abstract projects and hopes to find value with built cases examined through different views.

It is important to distinguish the particularities of the structural engineering I refer to here. Structural engineering needs some connection with mathematics, calculation, and analysis. For some engineers that is enough to generate a creative process, while others believe in the "grace of construction"—essentially the idea of designing the optimum structure and making it manifest in the design. Some prefer to be "structural artists," as first popularized by David Billington when he said, "My first objective is to define the new art form and show that since the eighteenth century some engineers have consciously practiced this art, and that numerous engineering artists were creating such works in the contemporary world of the twentieth century."[3] But the subject we are interested in is how engineering can push the boundaries of architecture rather than its own limits; that means supporting architects and helping them understand how to get the best of their engineers. Purely engineering-led solutions, such as imposing a convenient geometry, have to be shunned in favor of opening up the concept of engineering rationality to embrace support for an aesthetic principle. The relationship between architect and engineer, historically, has always produced new and transformative work, but is it always good work? Or indeed is good work only produced when such a relationship exists? What is the role of the client and the contractors? The projects studied here reflect a variety of positions taken by engineers in working with "design" to develop a space for interdisciplinary discourse without necessarily becoming "service models" to the architects' conception. The way architects should react to this evolving relationship, developing new pedagogical and practice models, has been part of the focus of the research. Knowledge of the engineer's role in design can produce benefits to the project.

In the most "interdisciplinary" interaction, the conventions of one or both disciplines are expanded and improved through the encounter. But as both disciplines are professional, as opposed to purely academic, any "improvement" has a social dimension. That dimension will often take the form of the clients' and users' goals for a project (which are themselves distillations of social and economic factors). So the collaboration that makes up design invokes triangulation with the social context. This is one way of objectifying and calibrating the relationship between the two disciplines and making outcomes assessable in ways that avoid the merely self-referential.

The violent economic swings of the last four years—since the course first ran—demonstrate the range of ways in which the two disciplines can interact, and how external forces can calibrate these possibilities. The research has been opportunistic in taking advantage of the insights that this situation offers in three specific ways: gauging implications for teaching; exploring whether it is possible to design a way out of recession; and acknowledging the different answers recession demands in reply to questions about sustainability and responsible use of resources.

Teaching
The discourse aims to be a direct preparation for practice. We try to reveal the false dichotomy between traditional and new models of practice—the latter enabled by advances in technology—using a series of carefully selected projects studied by the class to produce a broad intellectual and educational impact. It is clear that today more (and longer) study is needed to cope with the advanced knowledge available to designers; past models of study are not always as relevant.

One could argue that students in architecture or engineering schools should focus their education on the confined boundaries of their disciplines, and that interdisciplinary thinking should be learned afterwards, as they begin to practice. We explicitly reject this proposition that students should not be exposed to interdisciplinary thinking, and we distinguish being interdisciplinary from being a generalist. During the research, project architects and engineers contribute to the teaching program allowing, for instance, discussions of 1) the approach of each architect, 2) differences between architects with a given method, and 3) the productive interplay between projects and architects.

Technology has deeply affected the character of the work in the period we study and plays an important part in the teaching method in two ways. First, new software increasingly combines engineering-led analytical capabilities with visual representation. Many of the projects discussed would be almost impossible to deliver without recent developments in information technology and computer-aided design, and their relative accessibility gives students insight into how projects unfolded. This interplay of technology with core ideas of thinking, explanation, reasoning, finding things out, questioning, and taking content and evidence from real projects is developed through credible authority rather than a patronizing authoritarianism.

Design and Technology
Engineering is about making the best use of materials or other resources. Understanding how materials perform in ever-increasing detail—in part through better techniques of analysis—and even developing new materials (or techniques that combine them) increase the range of possible solutions to any given problem. All this relies on sophisticated technology, research and development, analysis, testing, and modeling—which at times unlock unexpected characteristics or possibilities that may superficially be irrational, but in the rarified conditions of a particularly challenging project become rational. Technology acts here through its scientific, social, and symbolic dimensions.

But engineering also demands an appreciation of the context in which solutions can be assessed and the most promising ones graded to pick an optimum. Mediating in this way between the modalities of physics and economics suggests that innovative engineering could indeed help to adapt construction proposals to recessionary conditions and might even contribute to changing those conditions.

Another dimension comes into play when engineering aspires to contribute to design. The duality between physics and economics expands to include aesthetics and visual culture. This is not the place to trace the fraught but very real relationship between culture and economics, but John Kenneth Galbraith's famous dictum that "In the affluent society, no sharp distinction can be made between luxuries and necessaries" implies how problematic it can be for an engineer to relate scarcity to necessity, and abundance to redundancy.[4] Where Galbraith placed such decisions in an economic and social context, design-oriented engineering integrates them with aesthetic thinking.

When engineering merges with architecture in a design concept, the myriad and sometimes surprising possibilities that arise through technological thinking engender advanced aesthetic ideas. Saying that it is impossible to determine where engineering starts and architecture stops may sound like a cliché, but a project such as Phaeno or Highcross makes the observation real. The results came from a genuine interaction between two disciplines, each prepared to absorb the ideas of the other to achieve something new. The origin of such projects has sometimes been in a teaching environment.

Affluence, Scarcity, and Resources
Recessions fit with sustainability to the extent that they align economic and environmental parameters. Boom-generated explorations of possibility give way to responsibility, with a sharper focus on resource use. Again, the collaborations discussed here seek to expand the duality between economic and environmental factors by considering social and aesthetic ones as well. The outcome emerges from a field of contingent factors that an iterative process consolidates into a real physical object, moving beyond superficial "greenwash." In this sense it becomes difficult, potentially meaningless, and sometimes impossible to attribute particular design features to environmental or economic priorities. Such goals may well have been clearly stated at the start, but each iteration will have contributed to a process of evolution that make any linear sequence from ambition to result extremely hard to trace.

An underlying theme of the discussions presented here has been to demonstrate how the context of the project, rather than abstract mathematical principles, determines the rationality of the engineering solution. It would be absurd to apply a structural system devised for the flowing forms of a Zaha Hadid's Phaeno Center to the orthogonal geometries of

Ravensbourne College of Art and Design by Foreign Office Architects. Each aesthetic demands its own contingent and particular approach—even though at root both seek to create forms, spaces, textures, and experiences through the manipulation of light and matter.

Having a wide range of project types and architects increases the richness of conversations and prevents the discourse from becoming a school for a certain "taste" or the response of one type of engineering attitude. But it also introduces a critical edge into the discussions, as each design approach can be challenged, complimented, or criticized. Furthermore, by following the project from conceptual design through fabrication and construction, the relationship between design and making becomes apparent. Beyond that, it shows how the logically convergent focus of structural engineers can interact with the power of imagination that architecture often promotes. This interplay is necessary to find routes out of the individual challenges that these disciplines face today in overcoming some lack of confidence from those we serve.

Inseparability of Strategic, Detail, and Procurement Issues in Design
One group of case studies shows two extreme positions: the UK Pavilion at the 2010 Shanghai Expo, designed by Thomas Heatherwick, and a couple of housing projects: the student dormitory for Queen Mary College of London University, by Feilden Clegg Bradley, and the Adelaide Wharf "affordable" housing in London, by Allford Hall Monaghan Morris.

In the temporary UK Pavilion, exploration is overt. It was a box suspended in a field of thousands of fiber-optic cables, which had to contain displays and provide space and facilities for visitors. Its structure was deliberately conceived as a spectacle, creating surprise and flouting convention, but its functional purpose prevented it from being a pure piece of sculpture. The challenges were numerous and only resolvable through extensive computational analysis and digital fabrication.

These were different goals from those of the two housing projects. In one case architect, engineer, and contractor collaborated closely to find the most efficient way of building the repeated standard unit of a bedroom. It turned out to be tunnel-formed cast in situ concrete (an old technology), a technique that had fallen out of use in Britain but fitted the particular parameters of this project in performance, cost, and appearance. Here exploration may have led back to an established technique, but it found a solution that was better than conventional options.

Another pair of case studies gives insight into sustainable design. Heelis, also designed by Feilden Clegg Bradley, is the headquarters of the National Trust, a large charity devoted to preserving historic landscapes and buildings, located in Swindon, a town between London and Bristol; Masdar is an entirely new city in the Abu Dhabi desert, designed by Foster and Partners. Both set high standards for environmental performance, in Masdar's case to achieve net-zero carbon emissions, net-zero energy use, and net-zero waste while making the desert habitable and providing water by desalination.

There the similarities end. Heelis's appearance derives from the roof form of an adjacent nineteenth-century industrial shed, but with its trapezoidal shape synthesizing contemporary requirements for light and heat. On a tight budget, getting the performance of the roof right meant that the accommodation itself could be very simple, spread across two stories with frequent atria punching holes in the upper floor to provide daylight throughout—and with enough height to use natural convection for ventilation.

At Masdar, budget is less of a concern than demonstrating how the desert can be made habitable in a sustainable way. The first phase, now complete, is the Masdar Institute of Science and Technology. Though the structure is relatively simple, the orientation, shape, and surfaces of the buildings all have to drive down the amount of energy needed to keep them comfortable, enhancing whatever wind there is for cooling and where possible using photovoltaic panels to generate clean power. As the inaugural element of the new city, it has to function on its own, but also establish the principles of the construction methods and urban form of the buildings that will grow up alongside it.

The case studies in the research also give a perspective on projects that exploit information technology in analysis, representation, or both, and ones that are more conventional. Designs that are apparently simple in form and construction can also benefit from powerful analysis, while complex forms are not necessarily made with the most advanced methods of production. The Henderson Waves Footbridge in Singapore, for example, designed with George Legendre, used a parametric equation to define the form of the plan and section, which change continually along its 250-meter length, but the structure was built by more traditional means.

Conclusion

Above all, the starting point is aesthetic curiosity—to understand first what the visual intention is, and only then try to appreciate how it is engineered. For this reason the research is a work in progress; in our practice, Adams Kara Taylor will continue to work collaboratively with architects on new (and we hope) challenging designs that will stretch our imaginations and resources and allow more cases to be examined in the school environment. To assist this aim, our p.art research team is expanding, moving from researching solutions to particular projects to proactive development of principles and processes—but they will still need to interact with architectural concepts to give the work validity.

We are seeing that the professional familiarity of the instructors and the experience of the teaching project itself has limits that can be transcended by the research each year. Students' papers and discussions clearly indicate that issues raised by such an experiment are critical to the development of future educators and practitioners. The best examples of the outcomes provide some answers, but also raise new questions for students, teachers, and practitioners.

Design is a visual discipline, and so innovation in design comes from visual stimuli; it is in the visual, rather than the digital or theoretical, that design engineering lies. Though the emphasis of the course is on design and engineering, the starting point is always in architecture and the aesthetic pleasure of the project. As an engineer, I have learned to appreciate this without shame—because that outcome validates the purpose of each profession that contributes to the design of buildings. It is in achieving it that design engineering defines its purpose.

Notes

1. Andrew Saint, *The Architect and Engineer: A Study in Sibling Rivalry* (New Haven: Yale University Press, 2008), p. 2.
2. George Orwell, "Politics and the English Language," *Horizon* (April 1946), pp. 252–265.
3. David P. Billington, *The Tower and the Bridge: The New Art of Structural Engineering* (Princeton: Princeton University Press, 1985), p. 4.
4. John Kenneth Galbraith, *The Affluent Society* (New York: Houghton Mifflin Company, 1958), p. 228.

Selected Readings

Antoine Picon, *Digital Culture in Architecture* (Basel: Birkhäuser, 2010).
Peter Rice, *An Engineer Imagines* (London: Ellipsis London Ltd., 1996).
Stephen Wolfram, *A New Kind of Science* (Champaign, IL: Wolfram Media Inc., 2002).

Learning from Design

Andreas Georgoulias

Lessons from design occur constantly during the process of articulation of specific solutions to given problems. Designers apply their tools to analytic, synthetic, and creative processes and, through constant feedback loops, arrive at their optimal solutions. As designs evolve and more projects come along, several patterns occur and their observation and documentation provide information about a given preference or approach. In interdisciplinary design, multiple disciplines integrate and overlay their knowledge, creative and analytic capabilities, and preferred approaches to deriving solutions.

The Harvard Graduate School of Design's recurring course GSD6328 started in the Spring semester of 2006 as an experimental exploration of interdisciplinary design. Its subject is design thinking that, whether a priori or impromptu, works within the interfaces, overlaps, seams, or gaps between the disciplinary areas of architecture and engineering. Hence the course title: "In Search of Design through Engineers." The name, however, does not tell the whole story. Our goal is always to look for design solutions, but the search is never through engineers or engineering only. Both among the instructors and, most important, among the students there is a mix of architects and engineers. Taking advantage the proximity of the GSD and MIT's Department of Structural Engineering, we market the course to students from both institutions. Our intention in having both engineers and architects in the classroom is to get as close as possible to a real design team. Sometimes the numbers balance, sometimes we have more architects in the class, sometimes more engineers.

Designing Collaborative Processes

Our approach to discussion topics, class materials, and assignments follows a similar path. We tend to opt more for messy realism than for disciplinary clarity through abstraction. We always start from the class materials. Drawing from Hanif Kara's vast resource of projects, key stage reports are given "as is" to the class. We take pride in the fact that a whole design report is made available, revealing unpublished details about projects that may have seen the light of publicity several times. For a two-week-long case study we give three, even four, design reports, representing the various phases of project delivery. That can amount to more than five hundred pages of material per project. Some students find this disorienting: "Not everything that is given to us is required to solve the assignment," one person told me in 2009; "it doesn't make sense, I don't know what to do."

Being the junior instructor in the class makes it easier to receive this type of feedback. However, addressing such comments is not easy at all.

They force us to think deeply and question our assumptions. At one point it became apparent that certain pedagogical approaches had led people to expect that every decision point should come in a package with its solving elements and a "how-to" guide to applying them. That desire for neat answers conflicts with our goal of getting as close as possible to the real design process and prepare students for the professional field. We believe that one needs not only to be able to distinguish the important from the unimportant but to prioritize among the available data and generate something from nothing to create one's vision.

Projects as Vehicles for Interdisciplinary Learning

The assignments are one of our principal vehicles to guide student thinking. Articulation, composition, and the requirements of the assignments become crucial for what we receive as an outcome and what the students learn in the process of delivering it. In that process, we also learn from them. At the beginning of the course, we were issuing more descriptive, more general, and more procedural tasks: "Which material(s) can you change to better reflect the architects' concept, and how will you deal with the cost increase coming from this material change?" Such phrasing of the assignment, we experienced, may be too generic and can result in weak analytics and less exciting results. The early assignments didn't necessarily force the students to arrive at a decision or propose a design solution. We reacted by simplifying and specifying what we asked for, by asking for design solutions, and by making it more radical: "Which two of the columns that support the Phaeno Science Center would you remove and why? What are the design implications of such a removal?" or "Redesign the Masdar Institute of Science and Technology if we move it from Abu Dhabi to Copenhagen" or "The sponsor of the Phaeno Science Center asks to double the building program while you are designing. What would be the implications on the structural system and how can you address this challenge?"

Such pointed assignments force each student to take tougher decisions, and since all assignments are done in teams, to decide as a team. The team dynamics become very interesting as we ask students to explain their reasoning and how they arrived at their solution, say, to removing columns 2 and 6. At that moment, many of the differences, similarities, synergies, and overlaps between architects and engineers would become apparent. Their starting points and means of reasoning sometimes coincide, but in many cases are different. Architecture students use more spatial and qualitative, even formal arguments. Several times they attempt logical leaps that lead to inspiring alternatives. "We propose to use the void that results from removing the two columns to bring light into the basement of the building—it's pretty dark as it is. And now that there is light in this uninhabited area, how about we change the floor material, insert some small plants and create a living space?" an architecture student would explain concerning their design development. Engineers will usually start from a purely quantitative analysis and employ their armature of software tools to demonstrate their argument: "Look how large the bridge deflection is!" an engineering student said

enthusiastically as he played an animation generated from the structural analysis software SAP.

However, a concern is that the two methods of reasoning can be independent and develop in parallel. As most of the assignments have several parts, each team tries to efficiently allocate resources, and the architects will tackle the design parts while the engineers take on the calculation parts. That pattern becomes obvious during class presentations, as each team member presents his or her part. If tasks are divided this way, several parts of the assignment will have little connection to the ones before and after, and the whole presentation will not deliver a unified message, nor a solid approach. To achieve more collaborative approaches, one has to be more radical. For us, that sometimes requires arranging the assignments in a way that determines who does what. At one point, it was tempting to start an assignment by asking the engineers to do the design part and the architects to do the engineering analyses. The teams actually enjoy this and the subsequent assignments, where we do not specify who does what, become more integrated and roles change more often. This is certainly a moment of gratification for us. Students end up presenting a part of the assignment in pairs, and the way they interchange their talking points and how they comment on what their classmate just said signal to us that they had worked together.

Unifying Patterns
Through the years, we constantly reflect on and learn from the outcomes of the class. Student assignments, final papers, course evaluations, and verbal feedback from students and guest speakers, all come to our review board. As we are eager to deliver on our promise of interdisciplinary pedagogy, we want to learn from these sources and improve our message and approach.

A common thread we find over the years is a set of trajectories, or themes, that run through the different assignments and case studies. Seemingly independent of a project's design approach or an architect's derivative position, these threads emerge every year and establish unifying patterns that we adapt to, learn from, and react to in different ways and intensities. One thing we cannot do, however, is ignore them.

These patterns reappear in material coming from us. For example, we use them in case-study topic matrices for the students to elaborate in their final papers. It is through the realization of these unifying patterns that, in 2010, we decide to significantly alter the course and replace the final paper with a final design project—one that, instead of being a reaction to a case study presented in class, is an original challenge for the students to respond to using their creative and analytical tools, alongside the themes of the class.

We use these unifying patterns to drive an interdisciplinary approach to design. We introduce them in this chapter, together with a limited but representative selection of student work that generated or expressed them. For each pattern, we use excerpts from student essays that frame, supplement, contradict, and nevertheless help define these thematic areas. It has to be noted that all designs presented here are outcomes that respond to a given case study and a specific assignment, done in one or two weeks by a team of students taking a full course load at the master's level. As such, each design presented here is probably the outcome of approximately thirty to sixty hours of work from a team of three to five individuals, dispersed across the two schools and multiple disciplines. In each case, only a few samples of student work are presented.

Structure and Design
The materiality of architecture has determined the blood relationship between design and engineering. Even in Étienne-Louis Boullée's utopian architecture, there is an instinctive thinking of material, gravity, and construction. People often talk about the iconic or geometric aspect of Boullée's Cénotaphe à Newton. But there are also structural and material indications in Boullée's proposal. In Boullée's age, when steel frame had not come into being, a masonry dome or vault was the most efficient way of achieving long span. The circular thick walls held the spherical dome in the center and resisted push-

ing forces toward its periphery. However, the large opening on the circular wall presented the conceptual or the modern part of the architecture, which was different from the Pantheon or the Colosseum. We can see that there is an interesting struggle and mutual influence between design and engineering even in utopian architecture. The engineering intuition somehow makes architecture grounded.

From Xiao Yin, "The Return of Engineering" (2010 student paper)

The grounding of structural engineering and architecture is the core of our first unifying pattern. The interplay between structure and design, the layering of processes and diverse but intertwined approaches to conceptualization, analysis, and construction emerge as architects and engineers delve deeper into their foundational base of knowledge and renegotiate the boundaries between their disciplines.

In the case study of the Henderson Waves Bridge in Singapore by IJP Corporation, Ilya Chistiakov, Isabel Lopes, Murat Mutlu, Scott Silverstein, and Hunter Young in 2009 revise the single-arch concept and introduce a double-arch solution. The team posits that a double-arch scheme generates a more porous structure, allowing the bridge to better blend with the surrounding landscape. This design alternative increases the deck stiffness and improves resistance to vibration, therefore making it possible to incorporate lighter support pylons. Finally, all arch members have a single curvature, making the fabrication process easier. The undulating aesthetics respond to this altered structural system in a similar fashion. A clear interplay unfolds between structure and form, and the team is agile in balancing the two in an optimal solution.

Things start to become more complex in the FOA projects. In the Shires project, the assignment is to redesign the glass façade system if it is not suspended from the roof, but rather attached to each concrete floor slab. Inherent in this challenge is the design problem of combining the differential deflections of concrete slabs and glass panes and steel trusses. Rosemarie Fang, Luis Aldana, Scott Silverstein, Murat Mutlu, and Mark Watabe in 2009 start with analytics and calculate the deflection of the

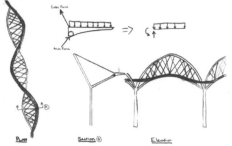

Sketches and variations for a twin-arch bridge concept.
I. Chistiakov, I. Lopes, M. Mutlu, S. Silverstein, H. Young (2009)

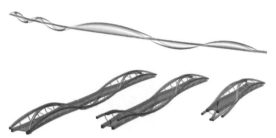

Rendering of a twin-arch bridge concept. I. Chistiakov,
I. Lopes, M. Mutlu, S. Silverstein, H. Young (2009)

Rendering of a twin-arch bridge concept. I. Chistiakov,
I. Lopes, M. Mutlu, S. Silverstein, H. Young (2009)

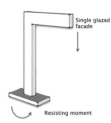
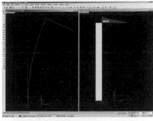

Gallows frame, bending moment analysis. R. Fang,
L. Aldana, S. Silverstein, M. Mutlu, M. Watabe (2009)

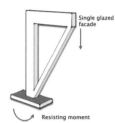

Adjusted gallows frame, bending moment analysis. R. Fang,
L. Aldana, S. Silverstein, M. Mutlu, M. Watabe (2009)

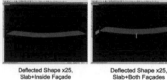

	Left	Center	Right
Concrete Alone	0.009m	0.022m	0.009m
Concrete + Inner Façade	0.015m	0.029m	0.015m
Concrete + Both Façades	0.021m	0.038m	0.021m

| Deflected Shape x25, Slab Only | Deflected Shape x25, Slab+Inside Façade | Deflected Shape x25, Slab+Both Façades |

Reinforced concrete slab deflection analysis. R. Fang,
L. Aldana, S. Silverstein, M. Mutlu, M. Watabe (2009)

gallows frames that support the curtain-wall façade, testing its reduction if they add a diagonal member and still support the façade without attaching it to the slabs. Then they calculate the concrete slab's deflection under different conditions of load. Their conclusion: the façade needs to be redesigned if it is to be attached at the slabs. The roof frame alteration won't suffice.

The team continues the façade redesign by proposing a connecting spider system that accommodates differential deflection. They proceed and profit from the new freedom that the new façade system offers, proposing a different ornate finish for the exterior skin—a woven steel cable in an Ammann bar grid.

The team responds to the core of the original architect's idea: a continuous pattern, visual porosity, and uniqueness. In their solution, structure and design combine after the support system of the façade changes.

Song He, Conway Pedron, YueYue Wang, Anthony Sullivan, and Jarrad Morgan in 2011 reach a design outcome of the same order by following a very different path. An extruded metal solution for the façade "laces" essentially transforms the pattern from a glass panel print to a structural, load-bearing element. The team, having secured their aesthetic position and fabrication technique, continues its analysis on pattern design and optimization in

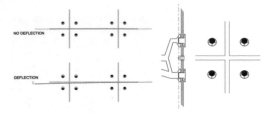

Façade design and detail for accommodation of slab.
R. Fang, L. Aldana, S. Silverstein, M. Mutlu, M. Watabe (2009)

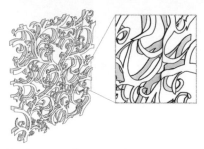

Façade concept where pattern becomes structural,
load-bearing, element. Design detail. S. He,
C. Pedron, Y. Wang, A. Sullivan, J. Morgan (2011)

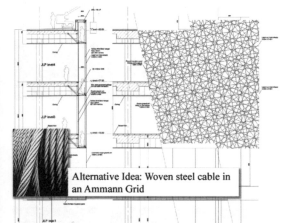

Alternative Idea: Woven steel cable in an Ammann Grid

Façade design for accommodation of slab deflection.
R. Fang, L. Aldana, S. Silverstein, M. Mutlu, M. Watabe (2009)

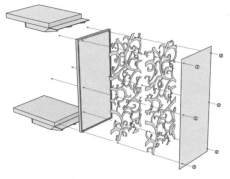

Façade concept where pattern becomes structural,
load-bearing, element. Assembly diagram. S. He,
C. Pedron, Y. Wang, A. Sullivan, J. Morgan (2011)

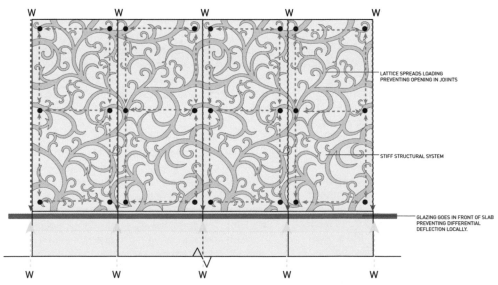

LATTICE SPREADS LOADING
PREVENTING OPENING IN JOIINTS

STIFF STRUCTURAL SYSTEM

GLAZING GOES IN FRONT OF SLAB
PREVENTING DIFFERENTIAL
DEFLECTION LOCALLY.

Façade concept where pattern becomes structural, load-bearing, element.
Front view. S. He, C. Pedron, Y. Wang, A. Sullivan, J. Morgan (2011)

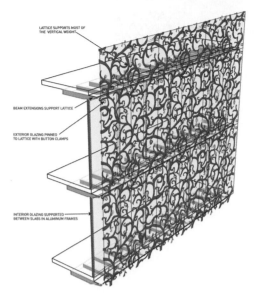

Façade concept where pattern becomes structural, load-bearing, element. Perspective view. S. He, C. Pedron, Y. Wang, A. Sullivan, J. Morgan (2011)

relation to deflection of slabs, light penetration and insulation, and buildability and cost. Again, structural considerations and design merge.

The Ravensbourne College of Art and Design, again by FOA, presents a different challenge: the dynamics of design and fabrication of a structural skin in relation to its façade pattern. The same design team, from the project above, now focuses on incorporating a structural diagrid at the façade, and compares it with a stepped column with a transfer-beam solution. They point out that coupling the diagrid with a flat slab allows for minimum structural depth of the slab and maximum flexibility of window placement. The diagrid, an inherently more efficient structural system, provides lateral stability and minimizes the need for concrete shear cores. The team concludes that a rectangular floor plan and removal of the stepped slabs allows for a consistent diagrid wrapping around the building. By connecting the diagrid to the slab always at a node, they simplify the connection details and reduce the actual amount of connections required, ending with a more cost-effective solution than a vertical column arrangement.

Ravensbourne College, structural system analysis with diagrid façade and comparison with stepped column alternative. S. He, C. Pedron, Y. Wang, A. Sullivan, J. Morgan (2011)

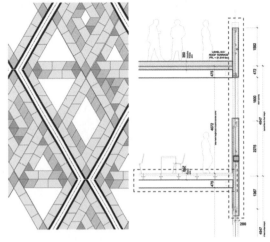

Diagrid façade section and detail. S. He, C. Pedron, Y. Wang, A. Sullivan, J. Morgan (2011)

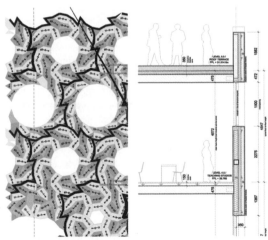

Diagrid façade section and detail. S. He, C. Pedron, Y. Wang, A. Sullivan, J. Morgan (2011)

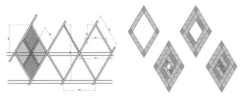

Diagrid façade pattern and tiling detail. S. He, C. Pedron, Y. Wang, A. Sullivan, J. Morgan (2011)

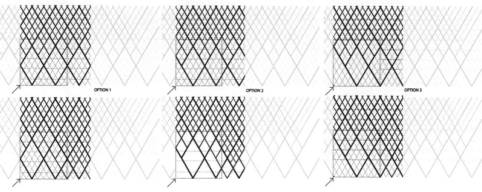

Ravensbourne College, diagrid façade geometric variations. H. Wu, L. McTague, Z. Wu (2011)

Ravensbourne college, diagrid façade tiling composition. H. Wu, L. McTague, Z. Wu (2011)

The team proceeds to study the façade pattern that emerges from their diagrid solution. Keeping the tiling solution of the original design, they alter the generative tile to fit a new façade geometry. The new triangular tile, distributed along the diagrid axes, generates a different system of composition to offer visual variety and conform to the building's varied programmatic needs.

Hailong Wu, Lesley McTague, and Zhou Wu in 2011 follow a slightly different approach in their analysis of the diagrid. They study primarily its geometric nature, as driven by floor heights and ground conditions, and evaluate the potential of different variations of a repetition canon. The team explores the interplay between primary structural members of the diagrid, the ones that lead the loads to the ground, and secondary members that populate a geometry of varying density.

The team applies this set of geometric and structural rules to the building's programmatic and volumetric requirements by creating a gentle and elaborate skin and a prominent entrance for the building's users. In this case, the structural nature of the diagrid recedes to a gentle mix of structural and nonstructural diagonal members, blending the boundaries of the load-bearing elements with the façade structure overlay. The typical boundary of structural skin and panelized glass façade blurs, with the drawback, perhaps, of material excess.

However, the key position of the team—a geometric approach to the diagrid—leads to a fabric-like wrap of the building that responds to local conditions such as the relationship with the ground and the entrance on mostly formal terms. On further elaboration, the design address structural loads through an optimization approach that maps physical forces and distributed loads to local conditions, and as a consequence defines the diagrid's pattern density.

The team continues their analysis and faces a new challenge: how to maintain a circular tiling pattern that FOA uses in their design, now that the structural canopy of the façade is diagonal. They perform

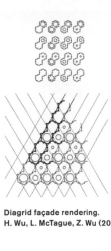

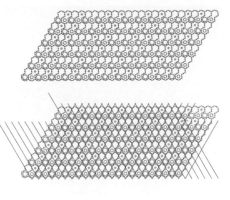

Repeating tile with varying aperture combinations

Variegated tiling edge makes integration with straight-edge diagrid members difficult to achieve

Diagrid façade rendering.
H. Wu, L. McTague, Z. Wu (2011)

Diagrid façade pattern.
H. Wu, L. McTague, Z. Wu (2011)

studies on design and composition of modular units, oscillating from elemental modules to macro-compositions to control the outcome on the key scales of the façade: the unit or tile, certain key local conditions (around a window for example), and the entire façade.

The team then tests the interaction of modular tiles with the diagonal axes of the diagrid. A compromise is bound to happen, as the rigid geometry of the diagrid disrupts a potential freedom of tiling composition, in terms of both resulting patterns and the actual design and assembly of tiles. Again, the relationship of structure and design is shaped by the principal thesis of this team: analyze the geometric nature of the diagrid and blur the boundaries of a structural skin with the exterior tiled envelope.

Based on that position, the moment where the pattern of tiles has to integrate with the structural skin is critical. Here, the structural geometry of the diagrid wins and the tile becomes rectangular, changing from a shape that could generate circular forms to a

shape that will generate rectangular and triangular forms. By careful elaboration and continuous design, the team maintains the rotation effect of the resulting pattern, without compromise of the integration of the structural skin with the tiled façade. An original thesis produced by an interdisciplinary team provides another successful demonstration of structural and design alignment. With a focus on the geometric nature of a diagrid, the façade and the structural skin of a building merge into a single system through a series of design decisions on the areas of structure, design, and aesthetics, as well as the boundaries between them.

Contextual and Systemic Design

FOA analyzed massing from a mainly programmatic standpoint. Some initial constraints were set, such as a connection to a carpark that could not obstruct heavily trafficked vehicular ways, and a relationship to an underground loading dock. While they conducted research on pedestrian and vehicular paths, their initial diagrams indicated an overarching concern for the program needs of their two

Phaeno center concept: remove two load-bearing cones. J. Bonner, M. Huang, L.S. Huang, J. Minguez (2008)

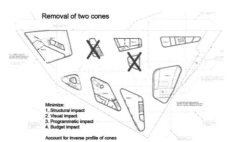

Phaeno center cone removal - ground floor. J. Bonner, M. Huang, L.S. Huang, J. Minguez (2008)

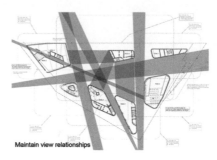

Phaeno center resulting view corridors at ground floor. J. Bonner, M. Huang, L.S. Huang, J. Minguez (2008)

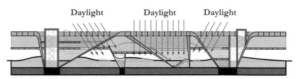

Phaeno center section after cone removal and resulting daylight penetration. F. Au, S. Leon, A. McGee, F. Mosqueira, M. Ruettinger (2010)

clients, the John Lewis department stores and the cinema owners. These colored diagrams map out potential adjacencies of the program, and all have certain commonalities. Among these are the large, uninterrupted green swaths that represent the John Lewis portion of the program. John Lewis had very stringent square footage requirements for continuity based on a tried-and-true spatial formula for attracting customers, and, as the anchor client, was loath to depart from these rules. So it was that John Lewis dictated a large chunk of the massing needs.

From Benjamin Lehrer, "Massing: Early Design Decisions with a Disproportionately Large Effect on a Building's Performance and Sustainability" (2010 student paper)

Design systems allocate resources in response to various factors and contextual constraints. Within a building, several apparent or implicit functions can coexist, with each system responding to a different set of variables. Optimal fit to context can be an inherently difficult design process when several systems overlay and vie for control of limited resources such as space, circulation, program, void, structure, energy, and light. Such dynamics emerge in the assignment given on the Phaeno Science Center, designed by Zaha Hadid. In this project, student teams have to remove two of the supporting cones and justify their selection.

Jennifer Bonner, Miaoyan Huang, Lee Su Huang, and Juan Minguez in 2008 analyze the structural support system of the building, looking for the two cones that have the least impact on the flow of gravity forces from the building to the ground. In other words, they start thinking like engineers. Immediately, however, the team sees that, in the absence of two cones, a different set of view corridors emerges at the ground level.

That alters their approach and leads them to evaluate the entire building in terms of views, light, and shadow. Their attention quickly shifts from the ground-level plan to the cross-section. The cones, instead of being structural members to be removed, become voids where light penetrates through the building in unpredictable ways. At this point, the team's system of thinking is changed.

This shift in systemic thinking leads to the entire reconceptualization of the interior experience of the building, by continuously penetrating the roof and slabs, as exemplified in the work of Fai Au, Sabrina Leon, Andrew McGee, Fernando Mosqueira, and Megan Ruettinger in 2010. The question is now how and where to bring light in, and not which two cones to remove. The system of design shifts from structural analysis to the relationship of the building's enclosed spaces with the outside. Openings function as light wells, and not as vacancies recalling a cone that once was there. More important, new meanings and opportunities emerge for the use of the space; the basement acquires a new program and function, becoming a place for relaxation and multiple activities. A vacant space before, now it is conceived as full of people and part of the building's activities. Finally, the quest that starts from the cones leads back to them, as the team changes the structural material from reinforced concrete to steel mesh, to better reflect and refract the light coming in. A systemic design shift had come full circle.

Similar systemic design changes, as adaptations to context, manifest themselves in the Heelis building case study, from FCB Studios. Travis Bost, Sophia Chang, Ji Seok Park, Andrew Pedron, and Danxi Zou in 2011 have to redesign the building if it moves from London to Abu Dhabi. The team reevaluates the building's sustainable design principles, function, and design elements. They start from the key functions of the building's environmental performance in its original UK context. Natural ventilation is a central aspect of the building's design, with the interplay of air flow and natural lighting being a major driver of shape, form, location, and dimension of openings, and general arrangement of solid versus void.

Phaeno Center interior before roof openings.
E. Kotsioris, A. Taylor, B. Pekoglu, K. Won (2010)

Phaeno Center interior after roof openings.
E. Kotsioris, A. Taylor, B. Pekoglu, K. Won (2010)

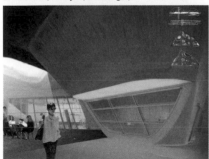

Phaeno Center ground floor after cone removal. J. Han, M. Holmquist, I. Lopes, A. Sambhare, H. Young (2009)

Phaeno Center ground floor with structural steel cone. F. Au, S. Leon, A. McGee, F. Mosqueira, M. Ruettinger (2010)

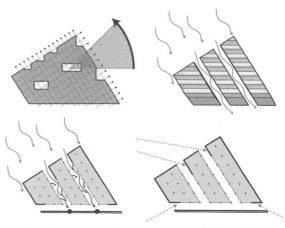

Heelis context and orientation analysis — Clockwise from top left: Heelis in Abu Dhabi; Prevailing winds and roof structure; Sunlight and floor structure; Prevailing winds and floor structure. T. Bost, S. Chang, J.S. Park, A. Pedron, D. Zou (2011)

Roof Adaptation

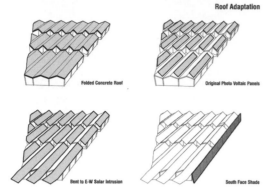

Folded Concrete Roof · Original Photo Voltaic Panels · Bent to E-W Solar Intrusion · South Face Shade

Heelis in Abu Dhabi. Roof structure and design. T. Bost, S. Chang, J.S. Park, A. Pedron, D. Zou (2011)

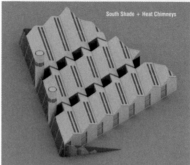

South Shade + Heat Chimneys

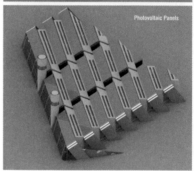

Photovoltaic Panels

Building design sequence

The relocation to a different context changes several of the climatic parameters that affect building design. First, the prevailing wind direction shifts, altering the relationship of building orientation and façade openings with the interior space and program distribution. The team is bold enough to change the entire concept and design of the building's envelope, openings, and atria, through a 30-degree shift in the design grid, to align with the new direction of prevailing winds in Abu Dhabi.

The shifting of the building's canopy has multiple consequences, in effect altering the layout of plans and spatial distribution of program. It creates major changes and generates several opportunities for the roof structure and design. Here one very interesting systemic change emerges. The openings stop being simply an entry point of natural light and ventilation; they become a defining part of the entire envelope design and structure, with very significant formal implications. A conventional undulating roof concept transforms to a much more complex system of panels, metal strips, and openings.

As expected, the systemic change in design does not stop at the building's envelope but continues to the core of its structural system. A light frame replaces the previous system of beams and columns, and aligns with the new roof structure and opening orientation. At the same time, the atria are transformed into very narrow, longitudinal canyons, optimizing the building's performance in the new extreme of climatic conditions.

The resulting building is quite different. The roof becomes a continuous envelope that wraps around the building in one move, as both roof and façade. The concept is refined through several formal variations within the system of panels, metal strips, and openings. New technologies that work in the new context, such as heat chimneys, are introduced, and existing ones, such as photovoltaic panels, are altered to exploit the new opportunities. Again, a series of systemic changes in design result from contextual adaptation. That, in effect, forever alters the building design.

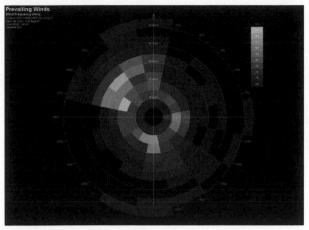

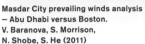

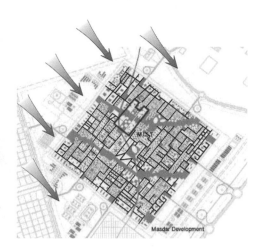

Masdar Development

Masdar City prevailing winds analysis
– Abu Dhabi versus Boston.
V. Baranova, S. Morrison,
N. Shobe, S. He (2011)

Another application of contextual and systemic design is the case-study assignment for Masdar Institute of Science and Technology (MIST), designed by Foster and Partners. For this project, Vera Baranova, Stephanie Morrison, Nathan Shobe, and Song He in 2011 perform the reverse assignment to the one given for the Heelis project: move MIST from Abu Dhabi to Boston and then maintain an equivalent environmental performance by reevaluating its design. The team starts from one of the key principles of design, the alignment of the city's grid with the prevailing wind direction. Through environmental and climate analysis, the team generates a new wind-rose diagram for Boston and subsequently rotates the primary grid to align with the new direction.

Continuing its analysis, the team presents simple diagrams to show the impact on building form and massing that the different context has in terms of sun angle, need for sunlight versus shade, and air flow. The wind is something to shield in Boston, rather than something to maximize exposure to as in Abu Dhabi.

These principles become inputs in the environmental analysis armature of the team, for tests of wind flow and solar radiance for various volumetric

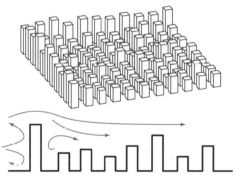

Volumetric wind analysis at city scale

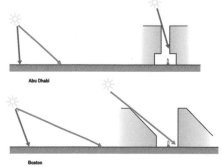

Abu Dhabi

Boston

Masdar City shading analysis – Abu Dhabi versus Boston

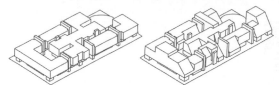

Masdar Institute volumetric variations – Abu Dhabi versus Boston.
V. Baranova, S. Morrison, N. Shobe, S. He (2011)

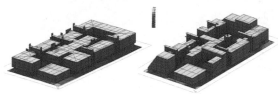

Masdar City solar radiation analysis – Abu Dhabi versus Boston.
V. Baranova, S. Morrison, N. Shobe, S. He (2011)

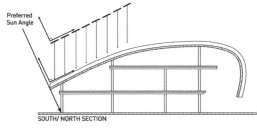

D

Masdar Institute auditorium shell roof section. Curvature and shape
variations, Abu Dhabi (top) versus Boston (bottom). V. Baranova,
S. Morrison, N. Shobe, S. He (2011)

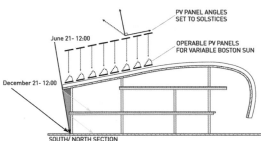

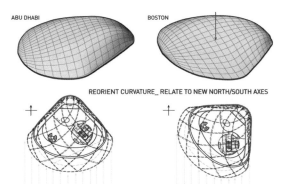

Masdar Institute auditorium shell roof plan. Curvature and shape
variations, Abu Dhabi versus Boston

alterations of the MIST building. The tool in this instance becomes an integral part of the design process, facilitating a feedback loop between environmental engineers and architects.

Moving to greater detail, Travis Bost, Sophia Chang, Ji Seok Park, Andrew Pedron, and Danxi Zou in 2011 apply the same principles of design to the main auditorium's shell roof. Instead of treating the shell's geometry as purely a formal function, the different sun angles become an input in the team's parametric analysis software. The curvature of the shell aligns with the new sun angle, while the entire structure rotates to adapt to the new context.

Computation and Design Optimization
The tools readily available to architects are a long way from being able to assess the performance of complex sustainable systems. The methods with which architects, and not engineers, can test such issues are incredibly limited. Many of the models have to be so drastically reduced in form and complexity that we end up testing something that no longer acts like the original design. We need to generate ways of easily testing new ideas and translating perfomative solutions found in biology into our own building practices. I do not think that our tools are ineffective or even inefficient; we just do not currently have ways of easily understanding how our ideas perform. If we cannot understand the way a boxfish moves with minimal effort though the water, we cannot then translate it into more efficient vehicles like the Mercedes-Benz "bionic" concept car.

From Megan Ruettinger, "Method: Investigations in Testing and Finding Appropriate Strategies in Biomimetic Sustainability" (2010 student paper)

Computation and design optimization frequently emerge in discussions and projects from the early years of teaching the course. Several tools, software programs, concepts, and approaches appear and reappear as teams strive to identify the optimal solution to a given problem. Sometimes the applications of the tools are the goal, other times the tools become just a means to a desired or unexpected outcome. We always direct the class toward the latter and urge the students to understand the risks embedded in the former.

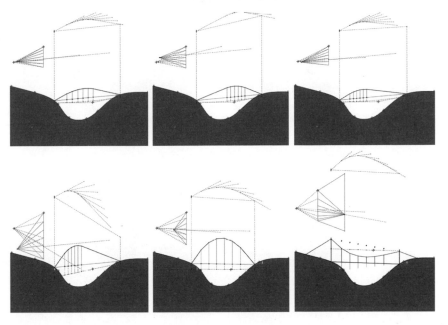

Bridge design through computational graphic statics method.
D. Papanikolaou (2008)

One of the most fertile, or maybe accessible, grounds for testing the interplay between computation and design optimization is IJP Corporation's Henderson Waves Bridge in Singapore. A clear and simple program (connect point A to point B) and a mathematics approach to design define the starting point of many decisions to come; eventually a bridge concept is born that is fascinating for architects and engineers alike. Dimitris Papanikolaou (MIT) in 2008 scripts an algorithm that uses the parametric graphic statics method to connect form-making and physics through a simple, geometrically driven application.

The algorithm can differentiate among several variations of structural systems and support conditions, and instantly generates the resulting form. This real-time interoperability is a critical dimension; both the design outcome and, more important, the ability of just one person to quickly generate this design algorithm are clear competitive advantages.

The search for design optimization through computation continues at the 2010 Shanghai EXPO UK Pavilion, designed by Heatherwick Studio. In this project, Rola Idris, Matias Imbern, Matthew Scarlett, and Patricia Semmler in 2012 have to solve

the pavilion for a spherical interior support shell, instead of the rectangular shape that was actually built. The design concept leaves no other choice for the team but to implement computational analysis for evaluation of design alternatives. Through Grasshopper, various combinations of exterior and interior attributes are modeled, controlling for the standard or variable length, and also for the single or multi-point orientation of the thousands of the pavilion's spikes.

The team continues to model the interior shell. They propose an egg-crate structure composed of metal ribs. All ribs aim toward the center of the sphere to create the least disruption to the placement of spikes. The isotropic nature of their solution provides ample challenges and opportunities to design the two polar nodes, each one having different load conditions, as well as hundreds of spikes passing through them.

Through careful computation analysis, five panel modules fit the geometry of the structural ribs; their repetition covers the entire area of the sphere. Assembly and manufacturing are the next steps of design, and the final effect of the British flag is evident, as in the original pavilion.

Parametric analysis of the 2010 Shanghai EXPO UK Pavilion, with original spikes orientation and envelope scheme. Grasshopper model, plan and section. R. Idris, M. Imbern, M. Scarlett, P. Semmler (2012)

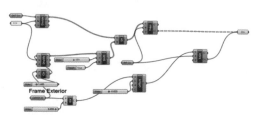

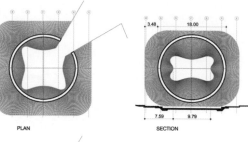

A. Current spikes envelope

PLAN SECTION

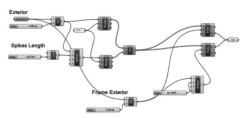

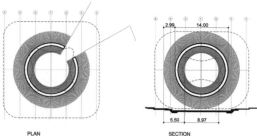

B. Spherical spikes envelope

PLAN SECTION

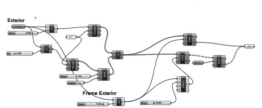

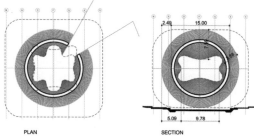

C. Mixed spikes envelope

PLAN SECTION

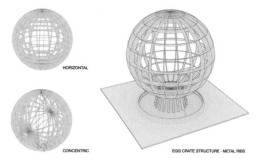

HORIZONTAL

CONCENTRIC

EGG CRATE STRUCTURE - METAL RIBS

Variations on interior shell structure

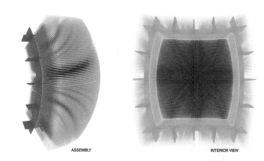

ASSEMBLY

INTERIOR VIEW

Elevation and front view of panel module with spikes

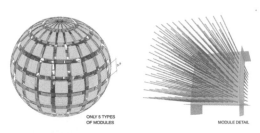

ONLY 5 TYPES OF MODULES

MODULE DETAIL

Interior shell panelization and module detail

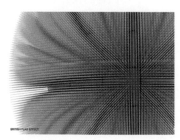

BRITISH FLAG EFFECT

Exterior closeup of the finished pavilion

On the same project, Ling Fan, Mariano Gomez Luque, Rebecca Hawton, Marianne Koch, Amaia Puras Ustarroz, and Bongjai Shin in 2012 follow a different path. The team does not embark immediately on modeling the pavilion and spikes in the digital space, but starts from a conceptual analysis of different stages of geometric transition from a cube to a sphere. A clever interplay between metaphors of known objects and their resulting topologies guides the team to transfer their ideas and generate alternatives. From sphere approximation, they move to resulting panelization and a subsequent evaluation on how many different panel types are needed to cover each tessellation step.

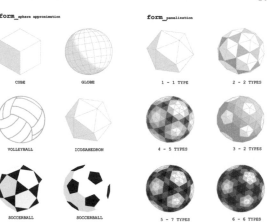

Geometric transformations of sphere: approximation and panelization.
L. Fan, M. Gomez, R. Hawton, M. Koch, A. Puras, B. Shin (2012)

The team continues its analysis on the relationship of the interior, structural sphere with the exterior sphere composed by the spikes, as well as the landscape in which the pavilion sits. They identify several alternatives of relative positioning of the two spheres, modifying along the way the density and orientation of the spikes. Concurrently, they examine the cross-section of the landscape podium and the spiked pavilion, proposing several deformation options for the optimal siting of the object. The relationship of the interior and exterior sphere, as well as the section of the landscape, generate several opportunities for structurally supporting the pavilion on its spikes, an intention of the original project that didn't get fulfilled.

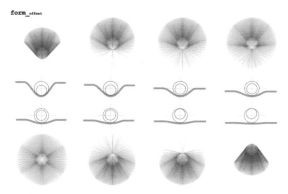

Variations on interior shell and spikes collocation, density and alignment

Through a combination of 3D modeling and rational thinking, the team derives its optimal design solution, covering the interior sphere geometry, the relative positioning of the interior and the exterior sphere that compose the pavilion, and the landscape formation that supports the pavilion. All the original design intentions are fulfilled through careful steps of design optimization and computation simulations.

Renderings of final output

The Emergence of a New Discipline

My interest, in particular, comes from a realization of the many difficulties and problems within the current mode of architectural design process that slow down and drain the power from the position of an architect. Architects are struggling to remain in control of the building design, with numerous contractors and consultants dictating their own terms and claiming a controlling position. Currently the profession has been outpaced and marginalized by a highly developed and technologically advanced set of specialists and subcontractors that maintain their expertise in some of the particular aspects of building design. While many of the adjacent trades have developed tools and methods of efficient and productive workflow, the majority of architects have not made this leap and fondly uphold traditional values and methods of design.

From Ilya Chistiakov, "Parametric Design + Engineering" (2008 student paper)

The designer's leading power, however, is not accomplished through superficial grandstanding with outrageous claims. One of the most effective ways to establish trust is through results. Success from past projects is an obvious plus, but this in itself may not be sufficient. Moreover, many firms when starting up, like IJP, barely have any projects to their name that they can leverage. It follows that more project specific results are necessary at the very start to truly win over clients. In each case presented in class, the architects were able to collaborate with AKT in a mutually advantageous way, by providing thorough development of innovative design choices when they were first suggested. This is a powerful technique: present a radical idea not as a whimsical musing but as a researched and genuinely realizable possibility. By wowing the other parties engaged in construction with a "source of superior knowledge" backed by earnest, project-tailored research, design firms gain respect and a competitive edge when determining project implementation and winning future clients' selection.

From Lester S. Yu, "Justifying the Unconventional, Giving Value to Design" (2010 student paper)

Too often in the design process, structural engineers are entrenched in the paradigm of converging on a solution given certain architectural constraints. This archetypal approach of discipline-specific problem definition quickly followed by solution optimization, while historically effective in many applications, can in another light hinder effective design that capitalizes on cross-disciplinary understanding. Hanif Kara and Andreas Georgoulias argue for the

"nonlinearization" of the design process through the fostering of this cross-disciplinary understanding. As the typical engineer who clings to quantitative problem solving, I often question what something as subjective as this really means. However, the mere question reveals to me a prevalent lack of cross-disciplinary understanding and subsequently validates the merits of a nonlinearized approach to design. At the very least, it paves a way for structural engineers such as me to attempt an answer to that question, "What is good design?" Furthermore, when applied to something as tangible as the future health of the planet, it is difficult for any engineer to deny stepping outside of the pervasive structural optimization shell. In doing so through this unlearning of typical linear practices, the challenges of design in the modern world will become more material and the ceiling for innovation will constantly rise.

From Hunter Young, "Techniques to Environmental Performance" (2010 student paper)

Interdisciplinary design is a function of simultaneous action around project-level and organization-level processes, values, and investments of relentless energy. Whereas the previous sections focused on the project-level analysis of what interdisciplinary design could do, this section will focus on what interdisciplinary practice could be. Several definitions exist, but the position derived is always the same: architecture and engineering disciplines need to reinvent the way they work together, both in terms of arriving at design solutions and also in ways of organizational collaboration and alignment. Determining borders between the disciplines that should not be crossed and boundaries that need to be renegotiated, the discourse incorporates notions such as nonlinear processes, productive naiveté, and unlearning.

Starting from the basics, Ioannis Ignatakis in 2008 maps the complex organizational structure of the Shires project. One of the useful lessons in this exercise is the realization of the multilayered dimensions of what a "project" is and what constitutes a "client." Such multiple levels of project organizations can lead decision-making into a complex and sometimes loosely controlled process. From simple design decisions to core design values and foundational project beliefs, the ability of a firm to navigate this complex interorganizational system is key in determining project success.

Murat Mutlu, Lana Potapova, Scott Silverstein, Jess Sundberg, and Mark Watabe in 2009 elaborate on such complex interorganizational systems through their analysis and procedural diagrams of a conventional design process versus what they identify as a contemporary design process. In these two diagrams one can see the linear approach of the conventional design process, and the fact that several decisions pass through or feed back to a single point in time. If a team gets this point wrong, the entire project is jeopardized. The very fact that this process is linear, and that several key project members such as the engineer enter the project at a later stage, only aggravate risks and uncertainties and increases costs of design changes.

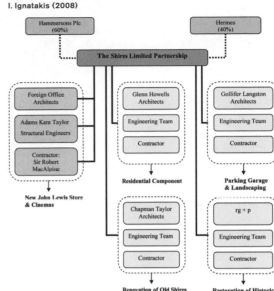

Shires development organizational scheme.
I. Ignatakis (2008)

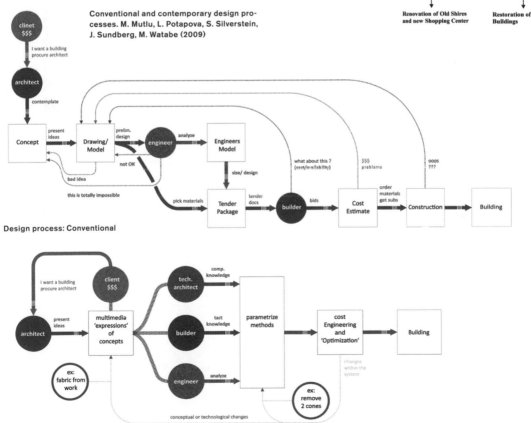

Conventional and contemporary design processes. M. Mutlu, L. Potapova, S. Silverstein, J. Sundberg, M. Watabe (2009)

Design process: Conventional

Design process: Contemporary

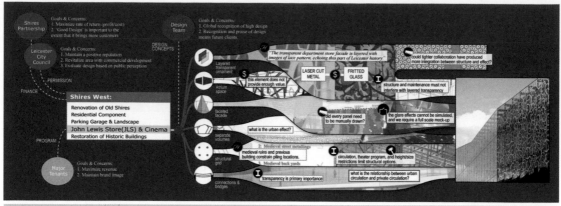

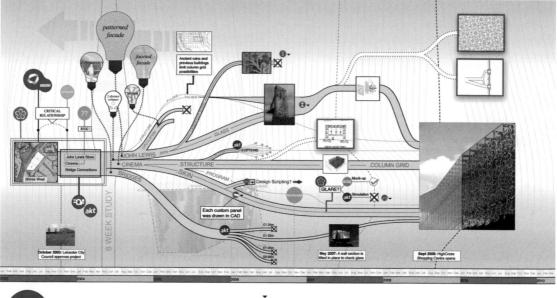

Shires development. Nonlinear design process diagram. M. Watabe (2010)

On the other hand, the contemporary design process diagram introduces a good example of nonlinear design. By a parallel development of multiple "expressions" of concepts from all key parties, the team demonstrates how interdisciplinary design can uncouple risks embedded in a linear and inflexible process, and demarcate roles and responsibilities in a highly collaborative formation. The role of technology and building information modeling in facilitating this process is evident in the identification of a single model, or depository of outcomes, from this multiple and interdisciplinary elaboration of design concepts.

The idea of multiple and differential expression of concepts are further cultivated by Mark Watabe in 2010. The evolution of a building's concept design is subdivided in different aspects of program, materials, structure, technology, etc.; in the diagram shown, each aspect, or expression of concept, is overlaid with all others in a constant interaction. Key moments in the delivery of the project are identified through the incorporation of icons and project images, showing the evolution of design through pivotal decision points.

Watabe further develops the visual language to express nonlinear design processes and interdisciplinary collaboration. Symbols and patterns express the primary idea generation, as well as the exploration of multiple concept alternatives. Design development is represented either as a straightforward line or an undulating path of discovery, amplification, or termination of ideas. Core design processes are differentiated from peripheral ones by a thicker line, to identify focal areas for each project. Escalation of a process line on the y axis signifies the escalation of project risk. As such, thicker and higher lines show a larger risk concentration on important project dimensions than do lighter and lower ones that represent secondary project dimensions.

The final diagrams have different representations of interdisciplinary design and nonlinear processes, as exemplified by case studies from the class. The Henderson Waves Bridge demonstrates the clarity of the central design concept as a straight line, with several design decisions on materials, manufacturing, and construction means and methods as parallel branches of concept implementation.

The emphasis on the design stage is repeated in the Queen Mary Student Housing project, where a simple and clear design brief underwent an important design optioneering phase to determine its construction method and its alignment with the strict budget and schedule constraints that would subsequently add risk to the project, as shown by a thick escalating line.

The Phaeno Science Center, on the other hand, demonstrates a continuous escalation of risk as the project unfolds; emphasis is on the construction phase, as several iterations on material properties, constructability, and manufacturing unfold from design throughout construction.

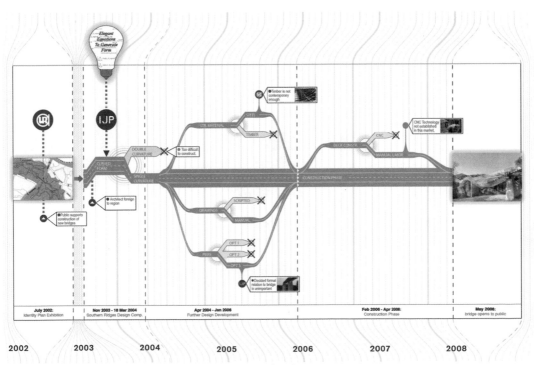

Henderson Waves bridge. Nonlinear design process diagram. M. Watabe (2010)

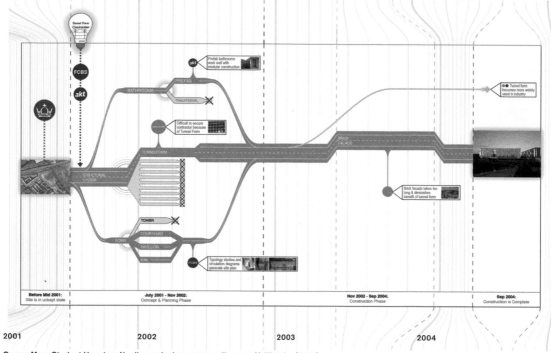

Queen Mary Student Housing. Nonlinear design process diagram. M. Watabe (2010)

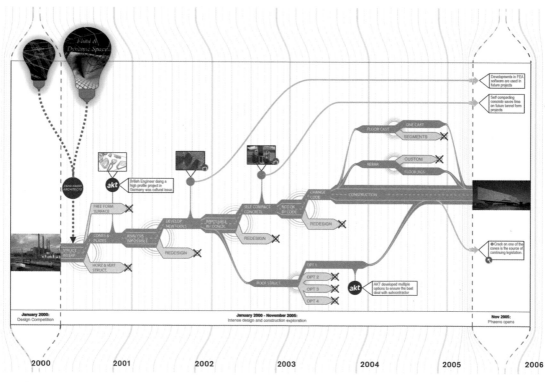

Phaeno Science Center. Nonlinear design process diagram. M. Watabe (2010)

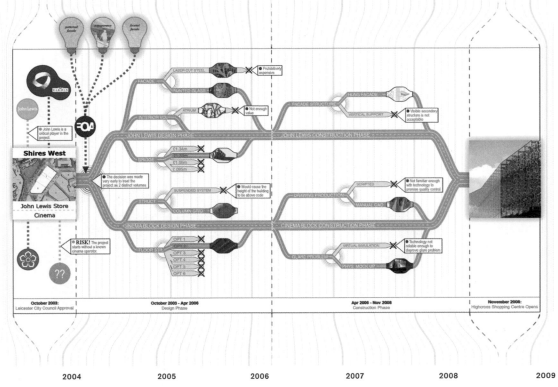

West Shires development. Nonlinear design process diagram. M. Watabe (2010)

Finally, the Shires project diagram represents the mixed-use dimension of the program, and the several design implications that a certain program, as well as a given client, would have on design evolution and alternative concept exploration.

Conclusion

The journey of interdisciplinary design, as studied at Harvard Graduate School of Design and exemplified through the case studies and assignments described within this chapter, doesn't end. We hope that this publication will help to clarify the issues, and provide operational ideas and methods to act upon. More important, it is our aim and expectation that these lessons from architecture and engineering may become an integral part of future design processes and lead to inspirational architecture and engineering, and more exciting built work. As practitioners and educators, it is our mission to incorporate the aspect of implementation within design—as messy,

mundane, or banal as that may sometimes be. But a project has to get built to be really amazing and benefit everyone.

Through the exploration of extremes, we strive to identify great ideas at both ends of the spectrum: the highly publicized projects, as well as those that were severely constrained by money, time, or circumstances. Wherever you find yourself in the field, we hope that you will keep crossing the borders and redefining the boundaries of our industry—that you will redefine what started as a specialization but ended as disciplinary fragmentation. In this publication we cannot provide all of the answers about where the dynamics may end up. But we pledge to maintain the dedication to our beliefs, act upon them, and keep researching the possibilities. Projects that now keep us busy will become tomorrow's educational avenues, and future explorations will lead to more questions and, we hope, more answers.

The interplay between structure and design, the layering of processes and diverse but intertwined approaches to conceptualization, analysis, and construction emerge as architects and engineers delve deeper into their foundational base of knowledge and renegotiate the boundaries between their disciplines.

Structure and Design

Structure as Architectural Intervention

Sabrina Leon

Structure is columnar, planar, or a combination of these which a designer can intentionally use to reinforce or realize ideas. (...) Columns, walls and beams can be thought of in terms of concepts of frequency, pattern, simplicity, regularity, randomness and complexity. As such, structure can be used to define space, create units, articulate circulation, suggest movement, or develop composition and modulations. In this way, it becomes inextricably linked to the very elements which create architecture, its quality and excitement.

—Angus J. MacDonald, *Structural Design for Architecture*, 32.

The viability of an architectural idea is largely dependent on structural design, despite being primarily evaluated in the context of aesthetics and emotion. Yet, the need to consider structural aesthetics as an inherent component in the creation of beautiful architecture continues to be a subject of debate. Structuralists maintain that a building's beauty follows from its structural efficiency; the more efficient the structure, the more beautiful the architecture. Most designers, however, disagree with this idea, believing that efficiency is an enemy of possibility.

When considering the influence of structure on architectural aesthetics, it is important to differentiate between buildings where the expression of its structure is relatively unimportant and those where it is essential. The Eiffel Tower, for example, capitalizes on the exposure of its structure, and its beauty is both iconic and transcendental because of it. In fact, the structure is the architecture. In other buildings, the structure is merely a framework mechanism, not part of the architectural image, and thus is purposely hidden from view. Most buildings, however, lie somewhere within this spectrum and have structural systems that influence the architecture. Thus structure plays a significant role in shaping the quality and experience of the architectural space.

Designing Structure for Architecture
Most designers understand structural design as a two-part process. The first part, or preliminary design stage, entails the selection of the form and the general layout of the structure. The second part is concerned with performing structural calculations and determining the size of the various structural members. In buildings that employ structure as part of the architecture many of the decisions regarding the design arise through the determination of the form. Here, the devised form dictates both the type of structure to be adopted as well as its material. Therefore the initial design concept, where form and the relationships of solid and void are determined, is a major controlling factor in structural design. If one wants to pursue the creation of architecture through the manipulation of structure, it is crucial to understand both the types of relationships that can be created between structure and architecture and the different ways in which structure produces or interacts with space.

Types of Relationships between Structure and Architecture
Some architects believe that a preoccupation with technical issues, such as those concerning structure, inhibits the creative aspect of design. Others feel that architecture should aim to resolve conflicts between form and structure and not allow one to

Exterior view of Waterloo station rail terminal, London.
Nicholas Grimshaw & Partners, 1993

Interior view of Waterloo station rail terminal, London.
Nicholas Grimshaw & Partners, 1993

dominate the other. This approach requires that the structural composition of a building be developed alongside all other aspects of its design; structural issues must be considered from an early stage and allowed to play a significant role in the determination of the final form. The nature of the relationship established between structure and architecture strongly influences the final form. These relationships can be catalogued as follows: ornamentation of structure, structure as ornament, structure as architecture, structure accepted as form generator, structure ignored, structure symbolized, structure and architecture as synthesized forms, and structure and architecture as contrasting forms.

Ornamentation of Structure

In this type of relationship, form is derived as an expressed structural armature, with minor visually motivated adjustments. Here the architecture tends to be tectonic, and the formal logic is largely a celebration and visual expression of the structural technology of the time. There is no attempt to disguise the structure or adopt forms that cannot be expressed by available materials; virtually everything visible is structural and technologically justified.

The train shed of the International Rail Terminal at Waterloo Station in London by Nicholas Grimshaw and Partners and Anthony Hunt Associates is a good example of this particular relationship. The arrangement of the exposed steel structure reconciles both aesthetic and technical considerations. In particular, the innovative use of a tapering steel substructure architecturalizes a new understanding of cross-sectional member efficiency.

Structure as Ornament

This relationship features a design process driven by visual rather than technical considerations, focusing on the visual quality of structural elements and attempting to give them visual prominence within the architecture. This approach pursues a visual agenda that expresses the structure to produce an image that celebrates the tectonic aspect of the architecture. It differs from ornamentation of structure in that these structures are often judged as less than ideal from a technical standpoint, as their performance objective is no longer solely structural.

The use of structure as ornament involves the creation of an unnecessary structural problem for the sake of visual drama. The process of finding an ingenious solution to the unnecessary problem is also the process that determines its architectural expression. Despite being visually interesting, many of the buildings that result from this approach are technically and structurally flawed. This is the case because the structures are expressed to convey the idea of technical excellence, but are not themselves necessarily examples of technology serving a specific structural function.

The Padre Pio Liturgical Hall in San Giovanni Rotondo, Italy, designed by Renzo Piano is one example where structure is used as ornament. The supporting structure is comprised of two interwoven rows of stone arches forming inner and outer rings. The arches of the outer ring are scaled-down versions of those on the inner ring, which originate at the center, where the altar is located. The spans of the arches steadily decrease, following the cochlear shape of the roof.

Interior view of Padre Pio church, Italy.
Renzo Piano, 2004

Palazzetto dello Sport, Rome, Italy.
Pier Luigi Nervi, 1957

L'Oceanografic Aquarium, Valencia, Spain.
Felix Candela, 2002

The visual drama created by the distorted shape of the arches, with shallower springing points at the center of the hall, produces the structural problem of force lines or thrust lines that deviate from the arches' cross-sections. To accommodate these force lines within the cross-section of each arch, as is required for structural stability, larger cross-sections had to be considered.

Structure as Architecture

In these cases, major decisions affecting the form are taken for structural reasons and are not compromised for visual effect. It is quite common for buildings in this category to approach the limits of what is structurally possible. Such is the case with long-span buildings, very tall buildings, and portable/temporary buildings. For long-span structures, the technical problem of maintaining a viable balance between the load carried and the weight of the structure itself dominates the form. Form-active typologies such as the compressive dome and vault, and the tensile membrane, are among the most structurally efficient examples for long spans. Despite their structural efficiency, these structures do have certain shortcomings. Their low mass makes them hard to design and build. In addition, they do not provide adequate thermal barriers, and their long-term durability, especially in the case of tensile membranes, is shorter than for most other building envelope types.

The works of Felix Candela, Pier Luigi Nervi, and Frei Otto are among the most notable in the long-span category. Nervi's invention of the two-way reinforced concrete space frame allowed for unprecedented thinness at great spans. Candela was notorious for the use of hyperbolic-paraboloid

Olympic Stadium, Munich, Germany.
Frei Otto, 1972

Sears Tower, Chicago.
SOM, 1973

Hearst Tower, New York.
Norman Foster, 2006

IBM Europe Travelling Pavilion, section.
Renzo Piano Building Workshop, 1986

geometries and thin reinforced concrete shells in his long-span structures. L'Oceanografic Restaurant showcases his mastery of this method. Cable network structures such as Otto's Munich stadium represent another method of achieving long span lightweight structures.

Tall buildings present other types of structural challenges, namely the ability to support themselves and resist high lateral forces such as wind. The first issue could be resolved by creating a strong connection between the base of the building and the ground. This has rarely been expressed architecturally. In fact, most buildings maintain a uniform size of the structure throughout each floor. The second issue has affected the architecture of the tall building. Most architects have chosen to place the supportive structure on the exterior of the buildings, allowing them to behave as framed or truss-tube vertical cantilevers. The Hearst Tower in New York and the Sears Tower in Chicago are both great examples of this technique. In the case of the Hearst Tower, the diagrid structure expressed on the exterior allows for a virtually column-free interior. The Sears Tower tackles the lateral loads via an internal cruciform arrangement of walls and columns. The form follows the arrangement of the internal structure as well as the basic structural principle of allocating most of the building's mass at the base.

The design of the lightweight/transportable building is heavily affected by the need to minimize weight and maximize assembly efficiency. This type of building is thus almost entirely determined by technical criteria. The realm of temporary buildings has been dominated by the tent as an efficient tensile mem-

brane structure. Compressive structures, such as Renzo Piano's IBM traveling exhibition building, have also been devised to serve this purpose.

The topic of portable architecture has even sparked the interest of the housing industry. In the summer of 2008, the Museum of Modern Art devoted an exhibition to the topic of portable prefabricated modern dwellings. "Home Delivery: Fabricating the Modern Dwelling" included Kieran Timberlake Associates' Cellophane House and O2 Village Architects' MicroHouse.

Cellophane House, MoMa, New York. KieranTimberlake Associates, 2008

Micro home 026 overlooking lake Zurich, Switzerland, Horden Cherry Lee Architects, 2008

Interior view of micro home at Berkeley Square London, Horden Cherry Lee Architects, 2008

44

Structure as Form Generator

Here structural requirements are allowed to influence the development of the form regardless of whether the structure itself is exposed. This type of architecture accommodates the adoption of the most sensible structural system amongst the options available for the building's particular scale. The intention is to produce architecture that gives equal relevance to all design aspects, technical and nontechnical. As with any work of architecture, the aesthetic and programmatic aspects are expected to be fully resolved, but the structural aspect is expected to be a stand-alone success and satisfy all technical criteria. In other words, though the structure may not be expressed, the design must accept its properties, requirements, and limitations as part of the visual vocabulary.

The early modern period produced many examples where the structure is accepted as a form-generating component. Le Corbusier's body of work is notable in this respect. As an advocate of form generated through structure, he frequently favored the use of the two-way spanning reinforced concrete cantilevering flat slabs over column grids. Corbusier's design for Villa Savoye and the monastery of La Tourette are just a few examples of his mastery of the reinforced concrete framework.

Modern skyscrapers constructed between the 1920s and 1930s also exemplify the adoption of structural technology as a form-shaping element. The form of the Empire State building, for example, is influenced by the steel frame structure and as such is representative of a successful marriage between architectural discourse and structural innovation.

Structure Ignored

Sometimes form is created without consideration of its structural implications. This was made especially possible during the twentieth century, with the use of construction materials such as steel and reinforced concrete. Vast formal possibilities arose from the materials' ability to resist tension, compression, and bending as well as from advancements in structural joinery. Several factors aligned

Villa Savoye, Poissy, France. Le Corbusier, 1928

La Tourette, Lyon, France. Le Corbusier, 1960

Empire State Building, New York. Shreve, Lamb and Harmon, 1931

Guggenheim Museum, Bilbao, Spain. Frank Ghery, 1997

Opera House, Sydney, Australia. Jorn Utzon, 1965

against the use of complex architectural forms, and twentieth-century architects availed themselves little of the potential for formal exploration. The most significant obstacle was cultural: modern ideologies favored orthogonality over curvilinear forms because it symbolized rationality; repetition and regularity because they mirrored the idealized assembly-line production theories of Fordism; and straight lines and sharp edges because they emulated mechanized and dehumanized production.

Another factor that played into the prolonged use of simple geometries was that of convenience. Orthogonal spatial arrangements were more suited to the standards of living of the time than were the implied irrationality of curvilinear forms and sharp angles. Cost was yet another factor that inhibited the use of irregular forms; complex forms were more difficult and therefore more expensive to construct. Still, the only technical obstacle was that of scale, brought on by the physical limitations of spans. Small scales allowed unlimited freedom in the matter of form, while larger scales and their implied larger spans faced structural constraints. Today scale is still the primary limiting factor on forms designed independently from structure.

Though an architect may choose to adopt a "structure ignored" approach to his or her design and devise the form in a purely sculptural way, the design process is still not entirely free. Two important considerations must be taken into account when form is devised without reference to structural requirements. First, the structure will likely have

to counteract large internal bending forces that result from implementing a non-form-active shape. Second, the internal bending forces are likely to be high in comparison to the load carried. These issues have design implications; the structural material will not be used efficiently, and the members will likely have large sizes. These issues are more persistent in larger spans, because they are more structurally restrictive and more expensive. As a result, the feasibility of the form is largely dependent on spans and their associated costs.

The introduction of the computer in the late twentieth century gave architects the freedom to design complex geometries. An increasingly powerful design tool, the computer enables structural analysis and simulation, and facilitates the description of complex geometries and the construction/ fabrication process. Computer-aided design opened a new chapter in formal exploration with works from architects such as Frank Gehry, Daniel Libeskind, and Jorn Utzon. Gehry's Guggenheim Museum in Bilbao, Spain, and Utzon's Sydney Opera House are representative of this type of architecture, where structural considerations had little to no influence on the original forms.

Structure Symbolized
In this category the structure is visually emphasized and allegorically represented. The architecture employs a visual vocabulary of images intended to convey the idea of progress and advanced technology. The origins of this vocabulary are for the most part non-architectural.

Lloyds Headquarters, London, UK
Foster Associates, 1986

Renault Distribution Centre, Swindon, UK
Foster Associates, 1983

In some cases the inspiration is rooted in science fiction but more often the approach borrows visual cues from activities with advanced technologies such as the aerospace, aeronautical and automobile industries, to create an image of technological progress. The resulting architectural style is dominated by forms and elements associated with structural efficiency. Structure is treated as a set of visual motifs, and the shape, size, and arrangement of structural elements is determined through consideration of both visual and technical criteria. The technical performance of the structure is secondary to the message conveyed by its image, and thus frequently compromised.

The high-tech architectural style, based on steel-frame structures, was one of the derivations of the structure symbolized approach. Steel was the chosen medium of expression because it was the only structural material that provided the desired image and a broad enough spectrum of component options for a substantial architectural vocabulary. Many examples of this trend exist throughout Europe; the Lloyd's headquarters building in London, for example, exhibits a visual vocabulary that is reminiscent of the aeronautical industry. In particular, aircraft structure-lightening techniques, such as cutting strategic holes throughout the structural elements to reduce their mass, are evidenced in the construction of the entrance canopy. The Renault Headquarters building in Swindon, UK, is another example of structure used as a central element in the building's image. As in other buildings of its kind, the complex structural system used here to reflect a company positioned at the forefront of technology and committed to quality is unnecessarily overdesigned and thus structurally inefficient.

The central problem in this relationship of design to structure is the degree of naiveté surrounding the decision to emulate advanced technologies for the sake of image. The technology developed in one field is not necessarily transferable to architecture, and therefore the image of progress is often symbolically misleading. For example, some advocate the use of this type of architecture maintaining that, because it appears to be technically advanced, it will somehow address worsening global environmental conditions.

However, as made clear by Angus MacDonald in *Structure and Architecture*, both the practice and the ideology behind the symbolic use of structure are fundamentally incompatible with the requirements of sustainable architecture. Symbolically articulated structures adopt images and forms from other technically advanced industries without assessing their technical suitability and are thus incapable of addressing the real issues surrounding sustainability. If an architect adopts the structure-symbolized approach in his or her design, it is essential that he or she remain true to the architectural statement intended. The structural form needs to be manipulated to make that statement obvious, or the approach will be futile and the architecture will appear unresolved.

Structure and Architecture as Synthesized Forms
In this variant, the structural system is ultimately responsible for defining the architectural form and envelope. This can be achieved through seven different structural systems.

Shell Structures: Shell structures or surface structures produce the purest integration of structural and architectural forms. They transfer loads within their material thicknesses and depend on their curvilinear geometry and support placement for structural performance. Shells typically define the roof form and thus act simultaneously as both structure and enclosure. They can be constructed from linear elements in steel or timber, but they are most commonly constructed of reinforced concrete, since it most efficiently resists compression and tension.

Fabric Structures: These structures function exclusively with tension. Like shells, fabric structures rely on three-dimensional curvatures for structural performance. Unlike shells, they require the assistance of supplemental compression members, such as masts and flying struts, to create points over which to stretch the fabric. Form, tensile strength, and thickness of fabric are determined for a set of expected loads. The interior illustrates the architectural qualities of the fabric and the compression members.

Ribbed Structures: Though they are capable of generating architectural form, these structures are frequently combined with a separate enclosure system. They are normally used to enclose single volumes as opposed to multi-story volumes, because multi-story volumes will likely require an additional internal load-bearing structure that would compromise the purity of the space enclosed by the ribs alone. The use of rib structures can be advantageous, as they can be strategically spaced to accommodate a degree of openness or transparency with the space.

Arches: Arches can be used to define the architectural form much like ribbed structures. Arches can be arranged in two directions to create a grid-like supportive structure for the envelope material. The framework created by the arches gives the finished envelope its final form.

Framed Structures: Frames are most commonly used to integrate orthogonal skeletal frameworks and rectilinear forms, but they have also been successfully used to integrate prismatic architectural forms. Frames can also be used in hierarchies such that the substructure is a composite scaled-down version of the main structure.

Walls: Walls can be used to integrate structural and architectural form. They can be used to dominate both the exterior and the interior of a building, and when used in rectilinear forms, they can strengthen the orthogonality of a design by supporting, enclosing, and subdividing the architectural space. Walls can also be used in combination with frames to articulate the space by establishing boundaries and implying circulation.

Structure and Architecture as Contrasting Forms
In this final relationship between architecture and structure, contrasts appear in the context of scale, geometry, material, or texture. There is an element of surprise typically associated with buildings of this kind; the structural form cannot be anticipated through a survey of the architectural form. Contrasting geometries between architectural and structural forms are evident at the TGV Station in Lille, France. The structural forms do not relate to the architectural qualities of the building. The curved roof does not rely, as one would expect, on a series of arched beams as in Calatrava's Guillemins Station, but on a lace-like steel truss system that doesn't follow the profile of the curvature.

TGV Station, Lille, France.
SNCF & Jean-Marie
Duthilleul, 1994

Guillemins Station, Liege, Belgium.
Santiago Calatrava, 2008

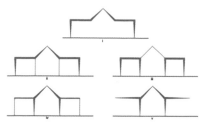

Structural layout versus space. After A. Ogg, (1987).
Architecture and Steel: The Australian Context, p. 49

Lyon School of Architecture, France. Perraudin/
Jourda Architectes, 1988

Structure and Space

Structural form and the way it is employed can have a tangible impact on space either enhancing or compromising the reading of the architectural idea. Alan Ogg's diagram of structural layouts and space provides a visual reference for how structure can influence the articulation of an interior space. The diagram provides a catalog of interior structural layout options for a particular envelope and their associated spatial qualities. The overall internal volume in options (i), (ii), can be perceived as one space with slightly different spatial qualities. Option (i) renders a continuous interior space, while option (ii) showcases a subtle subdivision of the space. Options (iii) and (iv), on the other hand, divide the space into two separate and distinctly different spatial zones, and option (v) pursues a closer relationship between interior and exterior.

Structure can thus modulate space in many different ways. It can be used to enhance or disrupt the building's function, maximize functional flexibility, articulate circulation, modulate or introduce light, organize, subdivide, or layer space, and create a phenomenological effect.

Structure for Enhancement or Disruption of Function

Interior structure can be used to enhance or disrupt a building's programmatic function. Structures can create a strong sense of habitation through modulations in density and scale. When used effectively, structure can help define the nature of a space. The glue-laminated timber struts at the Lyons School of Architecture provide a good example. The rhythm, scale, and arrangement of these members create

thoughtfully integrated human-scale spaces within a single volume. The school is arranged linearly with a central corridor of mezzanine work spaces flanked by double-height first-floor studios. The struts that support the roof also support the hanging work spaces on the mezzanine level via tension ties. The diversity of spaces created by the structure produces an environment that is both playful and intimate. The structure enhances the building's function by qualifying the overall space and framing the activities within the studios.

Structure can also disrupt the function of a building but this is rarely a deliberate intention of the architect. The seemingly random arrangement of columns in the Academy of Fine Arts in Munich, Germany, can be disruptive in the articulation and orientation of space.

Academy of Fine Arts in Munich, Germany
Coop Himmel(b)lau, 2005

Academy of Fine Arts in Munich, Germany
Coop Himmel(b)lau, 2005

O-14 building,
Dubai, UAE. Reiser
& Umemoto, 2009

Structure for Maximizing Functional Flexibility

Minimizing the amount of interior structure inside a building can be difficult, especially at large scales. Creating an environment that is clear of interior structure will maximize the level of flexibility with which the interior space can be organized and lead to more functional options. This requires the positioning of the primary load-bearing structure at the building's perimeter. In some cases this may not be a viable option given its implications on member depths and sizes, costs and spanning restrictions. Such reasons have led architects to favor the adoption of the free plan that allows spaces to flow freely through column systems, instead of being enclosed by load-bearing walls.

Other than structural shells, which are self-supporting, perimeter structures can be exoskeletal or endoskeletal. In exoskeletal structures, all structural members are located outside the building's envelope, while in the case of endoskeletal structures, they are contained within the building's envelope and thus restrict the continuity of the interior space and the functionality at the perimeter. Exoskeletal structures have been a topic of much fascination, particularly in the context of modern skyscrapers and office buildings. At 335 feet, the O-14 office building, completed in Dubai in 2009, offers a good example. Some of O-14's predecessors include SOM's 1961 Hartford Building and the 1970 John Hancock Center in Chicago. Unlike the more conventional Hartford and John Hancock buildings, with their Vierendeel truss and diagrid exoskeletal structures respectively, the O-14's unusual porous structural façade creates a completely column-free interior. The structure is composed of a self-consolidating concrete cast around a basket weave of reinforcing steel that was shaped around polystyrene masses to create the pattern of circular voids. O-14's façade structure further differentiates itself from its predecessors in that it also serves environmental functions. It addresses Dubai's high levels of incident solar radiation by providing solar shading through glazing, set a meter into the building to create a chimney-like cavity that facilitates surface cooling.

John Hancock Center, Chicago. SOM, 1970

Hartford Offices, Chicago. SOM, 1961

The degree to which an endoskeletal structure impinges on the functional flexibility of a building's floor plan depends on the designer. The structure can be incorporated unobtrusively, as when the structural depths and the envelope's thickness are sized equally, or when they are consciously integrated into the building's spatial or service functions. In the case of the Sainsbury Center for the Visual Arts in Norwich, England, the steel tube space trusses span columns of equal cross-section. Mechanical services, storage, and service areas are allocated within the 2.5 meter structural depth, allowing the interior space to function freely as an unobstructed public space.

Sainsbury Center, Norwich, UK. Exterior view. Foster and Partners, 1977

Colegio Teresiano, Barcelona, Spain. Antoni Gaudí, 1889

Sainsbury Center, Norwich, UK. Interior view. Foster and Partners, 1977

Structure for Articulation of Circulation

Columns, walls, and other structural elements have the ability to organize spaces and define or suggest the layout of the circulation. The type of structure used to define circulation, whether walls or columns, offers a direct means of restricting or directing movement. Column arrangements with tight interstitial spaces may restrict movement between spaces, much like a screen or any transparent partition. They may prevent you from relocating immediately but, by providing a glimpse of the adjacent space, influence how you choose to move through the building. Beams and other horizontal structures offer even more subtle suggestions of circulation routes. In addition, the scale, material, and rhythm of the structural elements can also help articulate circulation, particularly the emotional aspect of the transition between spaces.

The central corridor in the first floor of the Colegio Teresiano convent in Barcelona offers a great example of structure used to define circulation and evoke an emotional reaction. A sequence of white, pointed parabolic arches spaced 1.2 meters apart defines the seemingly interminable corridor that divides the classrooms on the first floor. The form, rhythm, and color of the arches interact with the soft, diffused light that filters through the light wells at the center of the space to create a serene and almost spiritual journey.

The roof structure over the Terminal 3 departure hall at the Hamburg Airport offers an example of structure used to enhance the direction of primary circulation. Against convention, the roof structure was designed to span the building's greatest dimension—its 101 meters length—to reinforce the landside-to-airside direction of passengers' circulation. The structure, composed of twelve curved steel space trusses, is supported on two rows of concrete piers spaced 61 meters apart via diagonal steel struts. The unconventional placement of the trusses between, instead of atop, the piers is used in combination with the glazed skylights positioned directly above to direct the flow of passenger traffic between members. The roof structure over terminal 2 at the Shanghai Pudong Airport is articulated with similar design motives. Pudong's trusses are also oriented in the direction of passenger traffic and supported between piers through diagonal struts. The intention of the placement of the truss and skylight system is to direct passengers between piers and across the terminal. However, unlike in Hamburg, these trusses do span the shortest dimension.

Terminal 3, Hamburg Airport. Von Gerkan, Marg & Partners, 1991.

Terminal 2, Pudong Airport, Shanghai.
Richard Rogers, 2008

Structure for Introduction or Modulation of Light

In *Structure as Architecture*, Andrew Charleson recalls Von Meiss' definition of space as something that exists when it is experienced by the senses. Structure plays a significant role in this context, especially when we consider its potential as a source and modulator of light. In fact, structure intervenes in the support of a building's envelope and roof. The degree to which structure inhibits or facilitates the induction of light is a function of both its configuration and its materiality.

Apart from its ability to function as a source of light, structure can also modulate light in three distinct ways. It can maximize light by minimizing its shadow, qualify it through reflection, diffusion, or screening, and utilize it to control its own perception. Some structural forms are more adept at allowing light through than others. Trusses and other lattice-like structures, for example, are quite porous, while solid structural walls are obviously not. However, in the majority of cases, most of the light that enters a space comes through the building's envelope or roof cladding via windows and skylights, and these are normally carefully positioned between structural members.

The roof of the Kimmel Center for the Performing Arts in Philadelphia is a case of structure used as a primary source of light as well as an example of its use to maximize light. The vaulted roof structure provides a transparent atrium that spans the entire area of the center. The structure is a combination of two systems, the barrel vault and the folded plate. Though the barrel vault form dominates the

Kimmel Center, Philadelphia.
Vinoly Architects, 2002

shape of the roof, the fine grain is detailed as a folded-plate structure. These structural systems accommodate the long spans and contribute to the roof's overall structural efficiency. The incorporation of the folded-plate action enables the Vierendeel trusses arching across the atrium to provide both vertical and lateral support, thereby resisting longitudinal wind loads. This structural strategy in combination with the use of a higher-stress steel material allowed engineers to minimize the member sizes, thus maximizing light. The structure's minimal interference with the light provides a unique spatial experience, allowing the space between the concert hall and theater to feel like an outdoor public square.

Detail of roof structure. Kimmel Center, Philadelphia.
Vinoly Architects, 2002

Detail of interior space between concert hall and theater. Kimmel Center, Philadelphia.
Vinoly Architects, 2002

Felix Nussbaum Museum, Germany. Daniel Libeskind, 1998

Interior hallway. Felix Nussbaum Museum, Germany. Daniel Libeskind, 1998

Monchengladbach Museum, Germany. Hans Hollein, 1982

The roof structures of the Monchengladbach Museum and Felix Nussbaum Museums both exemplify the use of structure to reflect and diffuse light. The white roof beams of the Monchengladbach Museum behave as a screen that filters direct sunlight and reflects it back into the gallery space, providing a soft, diffuse glow. The zig-zagging beams of the Felix Nussbaum corridor reflect and diffuse the artificial light to give the space a sepulchral glow—appropriate for a museum designed to showcase the work of a Jewish artist who died at Auschwitz in 1944.

Finally, the McNamara Alumni Center in Minnesota, Minneapolis by Antoine Predock provides an example of how structure can use light to alter its perception. Select solid wall planes are separated through a continuous glass strip ranging between 3 and 6 inches in height. The light penetrating through this strip dematerializes the wall plane by giving the illusion of weightlessness.

McNamara Alumni Center. Minnesota, Minneapolis. Antoine Predock, 2000

Structure for Organization, Subdivision, and Layering of Space

One of the most important roles of structure, aside from providing a building's supportive framework, is to organize, subdivide, and layer architectural space. Since the earliest works of architecture, structure, in its many expressions, has been used as an effective means of spatial intervention. It is as effective in creating subspaces within a large space as it is in defining their functional relationship. Structure provides designers with the opportunity to create a variation of spatial experiences within the architecture.

The Museum of Contemporary Art in Barcelona provides a good example of how structure can be used to layer space. Three layers of structure inhabit the building's atrium or ramp-hall. First, a layer of thin rectangular columns, supporting the roof and glazed skin, defines the atrium's outermost boundary. Next, a free-standing row of columns, alternating between nonstructural vertical elements, supports the ramping system cantilevering off of either side. Last, the atrium's innermost boundary is defined by the additional colonnade located in front of the beams and balconies emerging from the main galleries. Together, these structures produce a layered reading of the building's interior that is both evident through and echoed on the building's façade.

MACBA, Barcelona, Spain. Richard Meier, 1995

Wohlen School, Switzerland. Exterior view.
Santiago Calatrava, 1988

Stadttor, Dusseldorf, Germany. Petzinka Pink & Partner, 1998

Structure for Phenomenological Effect

Structures can also interact with space by pursuing a phenomenological effect. Most often this effect is produced through the sculptural treatment of the form, the use of an exaggerated scale, or both. The unconventional form of Calatrava's Wohlen High School hall in Switzerland, articulated through a self-supporting ribbed folded-timber vault and enclosing concrete walls, is representative of the sculptural and dynamic qualities that consistently dominate his work. The sculptural treatment of the concrete pedestals supporting the vault and the moiré effect achieved through the rhythm, orientation, and finish of the timber render a visually stunning interior space.

Structures can produce dramatic interior spaces through manipulation of scale. The visual drama within the Stadttor Building in Dusseldorf, Germany, for example, is produced by the exaggerated scale of the two concrete-filled steel-tube truss towers located at diagonally opposite ends of the floor plan. The tower members appear massive in cross-section, especially next to the building's columns, which appear frail in comparison. The towers, which serve as the lateral bracing structures of the building, run the full height of the building without any floor-slab interruptions and can thus be seen from virtually anywhere. The architect purposely chose to isolate the lateral load-carrying system from the gravity-supporting structure to express its resulting massive scale, as if trying to visually articulate the relationship between the building's loads.

Conclusion

Structure can be understood as a series of physical interventions used to support or execute a design concept. In this respect, the quality, rhythm, and complexity of the structural elements are powerful design outcomes that can be used to help articulate space, establish hierarchies, create thresholds, define circulation, modulate compositions, and, best of all, define experiences. These attributes represent a performative aspect of structural design that is uniquely architectural. Many designers do not avail themselves of the opportunity to explore structural design at this level for fear of creating architecture that is subservient to a chosen structural concept rather than an expression of the original design idea. When the structure is approached at the same time and in the same manner as the architecture and is motivated by the same design intentions, however the result is unequivocally stronger because the design concept is universally legible, profoundly integrated, and consistently applied. A comprehensively satisfying and aesthetically conscious solution to a design problem must recognize the potential for structure to enrich if not create architecture.

Bibliography

Allen, Isabel. *Structure as Design: 23 Projects that Wed Structure and Interior Design* (Gloucester, Mass.: Rockport Publishers, c2000).

Charleson, Andrew. *Structure as Architecture: A Source Book for Architects and Structural Engineers* (Oxford; Burlington, MA: Elsevier, Architectural Press, 2005).

Hertel, Heinrich. *Structure, Form, Movement* (New York: Reinhold, 1966).

MacDonald, Angus J. *Structure and Architecture* (Oxford; Woburn, Mass.: Architectural Press, 2001).

MacDonald, Angus J. *Structural Design for Architecture* (Oxford; Boston: Architectural Press, 1997).

Mostafavi, Mohsen (ed.) *Structure as Space: Engineering and Architecture in the Works of Jurg Conzett and His Partners* (London: Architectural Association, c2006).

Ogg, Alan. *Architecture and Steel: The Australian Context* (Red Hill, ACT, Australia: Royal Australian Institute of Architects, 1987).

Salvadori, Mario. *Structure in Architecture: The Building of Buildings* (Englewood Cliffs, N.J.: Prentice-Hall, c1986).

Tzonis, Alexander. *Movement, Structure, and the Work of Santiago Calatrava* (Basel; Boston: Birkhäuser, c1995).

Behind the Curtains: Backstage

Burak Pekoglu

An architectural project can be thought of as a theatrical production. A building becomes a stage after it is completed; those who build are actors who present the inhabitants–the audience–with the experience of the building through its very presence. The word architecture itself originates from the Greek αρχιτεκτων which means master builder, proposing a connection between the notion of the director in a play and its field of action, the theater. There is therefore an opportunity to define a relationship between theatrical production and architectural production in this light of establishing the architect as the director. The design idea is originally conceived by the architect, who directs the complex process of building, in which the contractor becomes an actor, and finally the end users serve as the audience of the architectural product. The site becomes a stage of hidden processes that determine the outcome of the design project.

Let us assume then that the architecture project, the building, could be perceived metaphorically as a play. The play could be staged at any particular location and actors, builders, and audience members multiply in relation to scale and theme. The project's effect can vary relative to the budget, theme, time, politics, place, and quality of its production. Most important is to make the audience believe in the performance. The building should perform in environmental, structural, economical, programmatic, contextual, and cultural dimensions. The outcome, as the performance, is determined by protagonists supported by the efforts of backstage teams, throughout the project. There is no particular way to orchestrate the production.

Accolades are awarded to those whose performance wins the applause of the audience. The relations and communications among them fluctuate in the spirit of the play. There is no formula that defines or records events in this multifaceted arena, where protagonists and backstage teams operate toward one coherent composition.

The play may be defined as preceding cinema in light of recent technological advances. Here we could introduce another metaphor by comparing stage and screen. The screen in cinema replaces the stage of the theater, as the art of drafting is replaced by digital drawing. The formats of these two methods of drawing are similar in their essence. Their output follows a similar language, where the speed of orchestration and experience of the audience are upgraded to vast visualizations through the medium of the screen. The projects are operated under new tools and communication techniques. Digital fabrication allows exploration of new means to translate composite surface designs to per formative material compositions. However, the backstage accommodates a series of different activities. Prior to the realization of the built product, the crew of architects, engineers, and contractors define core text and chapters of the play that is about to be staged. Here, learning and borrowing may come from the automobile, fashion, and aerospace industries. Although change is slow at the building industry, new lenses and screens propose a reconsideration of existing standards. Acquiring tools and technologies from other industries allows alternative possibilities of performance resolution to foster.

The last three decades have transformed the imagination of the designer into a more fluid and complex model of narration. Through the use of technological advancements, architecture has transitioned into the production of smooth and continuous surfaces. The cinema, having the ability of montage, is parallel to the limitless potentials of manipulating the surface in architecture through software. On screen editing becomes the way designers think as building culture advances. Data can be easily sent back and forth among all the participants in the theater of architecture. As processing speed is multiplied, it demands clusters of collaboration.

Editing can happen from a global scale to a local scale, until it reaches the audience as a final composition-image. The imagery that accompanies the new digital project of architecture does not necessarily display the reality of the architectural product. In this case, the spirit of the play is a mere suggestion of the intentions of the actors. The reality itself is not a static and defined image; it changes and transforms as influenced by the integration of these exact technological tools in the process of building.

The course "In Search of Design Through Engineers" given at the Harvard Graduate School of Design in the spring of 2010 evolved through six case studies, built architectural projects that present the engineering effort of Adams Kara Taylor (AKT). Interaction and collaboration between the architect and the engineer proposes a deeper analysis of the communicative potentials within contemporary practice. In this debate, the evaluation of traditional technologies of the recent past in relation to the production of contemporary design became a central theme.

The chosen projects were dissected in terms of their design conception and construction methodology, followed by careful analysis of the processes and technologies involved in their realization. This dissection demonstrates the temporal, climatic, political, and environmental contexts in which each of the projects were conceived. By re-designing certain components of these projects, one is led to a series of speculations on the opportunities for alternative solutions on given problems of planning, structure and infrastructure. Both the spirit of expression and the efficiency of construction were identified and used to inform the model of study. These studies address the challenges identified with concrete design and engineering proposals, establishing a model of practice in which the engineer becomes an active agent of design in the early stages of the architectural project, beside the architect.

Hammerson's Highcross Quarter Shopping Center by Foreign Office Architects reveals problems of the envelope, site strategy, and program in its realization. Understanding the decisions made in the sequence of design evolution is crucial to recasting or redirecting the whole play. The patterns of production in this case had to address the architects' aesthetic concerns, which resulted in textile becoming the concept for the envelope of the building rather than choosing to contain the mall as an opaque mass. The concept of the envelope manifested in the façade pattern. More precisely, this transformation took place in the process of exchange between the architect's fabrication design and the installation techniques devised by the engineer. Large components of the program, however, such as the adjacent movie theater, had to be resolved separately. The adoption of various systems and façade solutions caused programmatic and budget issues to be resolved under pressure. For better or worse, decisions had to be made fast given limits in time and budget.

In each project, design processes are unique in their constraints and paths toward risk taking. In the case of the Phaeno Science Center by Zaha Hadid, the ambitious stressing of the envelope challenges the continuous form in a redefinition of the monolithic structure that holds the building's programmatic and technical components. Here the idea of of form determinate in the architect's mind needs to be accompanied by the engineers' ability to meet the challenges. AKT frames such collaboration in its research potential of learning. The project can be conceived, as well as documented, as solving the problems of communication between the idea and the materialization of the building.

The orchestration of Phaeno takes place in Germany, a country dedicated to advanced manufacturing techniques and methodologies. The type of concrete used in the construction of the building, however, caused a significant delay due to the time necessary for its approval by the local authorities. Even though the advanced testing of concrete technology in the context of the project presented numerous benefits for the advancement of material science, in this case it seems that the choice of such material, with its inherent difficulties, presented a challenge despite the desired aesthetic effects. Yet there are undeniable benefits to sponsoring material research as a part of the architectural project. New challenges and potentials are brought to light with the extensive documentation of the backstage processes, which highlights the characters' beliefs in the play until they manage to stage its complete image. The audience is meant to be shocked and engaged in a truly new way. In addition to the initial exterior perception of the building as a monolithical sculptural object, the inhabitants discover a radically new expression of expansive spatiality, layered light structures, and materials in the interior of the Phaeno.

With the Henderson Waves Bridge by IJP Corporation, George Legendre opens up a debate concerning the stylistic approach of design and its implications for the process of construction. A small office in London won a competition in Singapore. The process deals with cultural disjunctions toward achieving the desired result. Legendre's specific approach to design, where math equations become a sketch for a tectonic form, raises an interesting set of problems in the consideration of aesthetic qualities and suggested structural functions. When the sketch model of the bridge translates into construction, gaps rise from lack of control. The architect's limited role during the construction phase results in unwanted alterations of the initial model. As a consequence, the director loses control over the production of the image. The

collaboration among the agencies becomes crucial to the project's implementation from concept to finished structure.

The Queen Mary Housing Project by Feilden Clegg Bradley is realized with the support of AKT. In this example, the tunnel form is revisited to address limits of time and budget. In this case, the approach was to tweak the units' organizations to gain architectural qualities of surface depth and dynamic over the façade. The tunnel form idea becomes the DNA for the units' aggregation within the limits of planning regulations and construction methods. Through façade additions to the north, rooms gain interest in their intrinsic layouts. The riverside block advances the image of the project in its alternative layout of unit windows and copper cladding.

Feilden Clegg Bradley architects challenge sustainability measures in their Heelis—National Trust headquarters project. Environmental analysis software becomes a tool for their intuitive design strategy. The skylight details and arrangement become the characteristic of the mat building, enhancing a monotonous building envelope while reflecting the directors' image by avoiding variations. In both projects, linear design thinking is prevalent, based on repetition of certain building elements with little variation.

The Adelaide Wharf Housing project by Allford Hall Monaghan Morris Architects demonstrates ways in which its courtyard typology is comfortably situated in its urban context. Potential dead spaces are activated in their organizational and spatial layouts. They enhance a safe access to the building, while trying to portray a social space. The bright colors add up to positive feelings in the audience, facilitating a closer connection with nature. In this scheme the strategies of architectural play become essential in the way they define the character of the project.

The social housing projects by Feilden Clegg Bradley architects and Allford Hall Monaghan Morris architects demonstrate the attachment to the live stage, where standard methods, manual labor, and linearity are still in control of architecture for economic and cultural reasons. These are two positive examples of how to rethink housing projects given their persistent limitations. Digital architecture excludes itself from this range of architectural production. Its presence is restricted to cultural and commercial projects, as it is apparent from the Phaeno and Shires examples. These tend to be more open to integrating the digital spectrum in the building process.

Editing is the key to arriving at creative architectural solutions. The cinematic thinking of being able to manipulate a single frame in relation to time and space could open up new possibilities for designers to think as creative directors. Speed matters in the way a director can alter design, increase the pace or slow it down, always responding to formal, spatial, and programmatic conjectures. The script is the framework, and the characters that control it can determine the spirit of the play, which is then experienced by the potential audience—the users who live in it. Within this scope, the complex process of architectural production reveals itself in the way it translates from a concept to built form, projecting one or multiple images. The image itself is translated into form, action and scenario, a trio that can describe the outcome of the built reality. In this light the architecture's capacity to transcend characters and capital becomes preeminent. In the final realization of architecture, projects are conceived as performative case studies when viewed as the experience and the documentation of the process itself.

Surfing the Wave

Murat Mutlu

When writing about Manhattan in *Delirious New York*, Rem Koolhaas suggested that the success of this city relied on the fact that its architecture had surrendered itself to the needs of the metropolis. This kind of architecture has the same relationship with the forces of contemporary trends as a surfer does with waves.[1] To follow the movements of reality is to synthesize observations from the real world in making design decisions. Without collaborating with actuality, the designer will get lost in irrelevant abstract visions. To surf the wave, any contemporary design practice needs to derive its aspirations from available opportunities, which requires a comprehensive knowledge of the constantly evolving market. Each opportunity—or hybridization of opportunities—becomes a design instrument with which the designer can develop ideas. Our skills as designers come from being able to design with what is already out there, rather than proposing ideas and forms that are derived from our fantasies of a controlled utopian world.[2]

Integrated Design Process

In the past, the "master builder" was able to comprehend all of the knowledge needed to construct an idea. When designing a building, he would know what materials needed to be used in what form, how the loads would be distributed in the structure, how the public would engage with the space. Because the knowledge necessary for designing the artifact was contained in one mind, the process of design was incorporated with these constraints of materiality from the outset. In the contemporary world, however, it is not possible

for one design practice or practitioner to have a meta-knowledge of construction technology, material science, structures, urbanism, information technology, and other fields required for a design to materialize. The industrial revolution introduced building materials such as iron that were new to the traditional master builder. To surf the wave of that time, the master builder/architect had to formulate his design knowledge about the new way of constructing forms by collaborating with experts. It is no coincidence that this was also when the structural engineering profession emerged.

Since the industrial revolution, in a traditional design process the architect will develop a formal concept of his or her design solution to the given problem that often lacks relevance to real material issues. It is not until the designer completes the concept that the building engineers start rationalizing the initial form. Among the critical avant-garde architecture practices, it was OMA that started working closely with an engineer, Cecil Balmond of Arup.[3] For them, the desire to explore opportunities led to collaborations with engineers knowledgeable about the potentials of materiality and also aware of industry-standard construction and fabrication techniques. This collaboration enabled the projects to be conceptualized with real material and construction issues taken into account from the beginning of the design process. For instance, in the Maison à Bordeaux project, this early collaboration enabled OMA to perforate the floating mega concrete beam—which also acts as a façade—to create windows for the rooms inside.[4]

Maison à Bordeaux, France. OMA and Arup

Meanwhile, engineers who have always been grounded in their approach to design in relation to reality established the discipline of design engineering, which creates and transforms ideas into a product that satisfies customers as well as business requirements. The role of the design engineer is the "creation, synthesis, iteration, optimization, and presentation of design solutions."[5] This textbook definition of the term can be interpreted as being a bridge between concept and product. The quality of the product relies on the strength of the concept; the strength of the concept relies on the quality of the product. The design process has to be well integrated such that the designer or the engineer (or the design engineer) is able to make decisions based on this reciprocity between concept and product—in other words, between ideas and reality.

The idea of collaboration with engineers at the conceptual stage of an architectural design process to create a better building is no different from the aspirations of design engineering. However, not until recently was the term "design engineering" made part of the mainstream intellectual discourse among architecture professionals. Hanif Kara of Adams Kara Taylor, for example, has defined the term as a creative design process of generating forms and organizations in which the available opportunities for both design and engineering constitute an integral component of architectural knowledge.[6] With the advent of sophisticated information technologies, the design process has shifted to become more flexible and malleable—no longer a linear and hierarchical information transfer between design and engineering. This convergence between disciplines has created an intimate relationship.

Phaeno Science Center, Wolfsburg, Germany.
Zaha Hadid Architects and AKT

It is now almost impossible to make a novel project materialize without this integrated design process.

One great example of such an integrated design process is the Phaeno Science Center project, the product of a successful collaboration between Zaha Hadid Architects and Adams Kara Taylor. The concept of uniting the severed urban fabric of the city of Wolfsburg and the architect's interest in fluidity of space gave birth to the design of a trapezoidal elevated volume that seamlessly blends with the ground, supported by ten inverted cones. The design breaks down orthodox definitions of what constitutes a wall or a floor. The cones and the waffle slab seamlessly join to make a continuous shell while performing the role of both structural elements (to support the upper floor and the roof) and architectural elements (to provide access to the upper level and accommodate programs). The waffle concrete slab and the roof pattern are articulated to adapt the trapezoidal plan—another example of the synergy between the spatial and the structural conception of the building. This symbiosis between

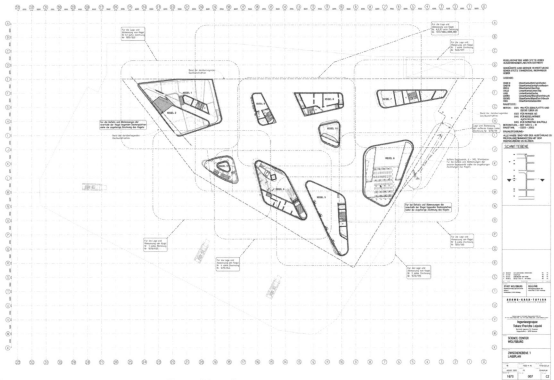

Ground Floor Plan showing cone structures. Phaeno Science Center, Wolfsburg, Germany. Zaha Hadid Architects and AKT

structure and space required a close working process between engineers and architects, as Patrik Schumacher, partner at ZHA, said: "The structure constitutes the architecture, and therefore the demand for tight collaboration was extraordinary."[7]

The architect's desire for fluidity and seamlessness requires the structural elements of the building to be analyzed as a whole system, in which any force affects all other elements, rather than as individual elements with their own forces, as in a conventional analysis method. There is no redundancy in the system. All the elements have to work together; the project was stable only when all ten concrete cones and the waffle floor slab were cast in place. The positioning of these cones was both an architectural and a structural challenge that needed to be solved by the two offices collaboratively. Material realities differently limit the span amounts of the waffle concrete floor and metal roof spans between

the cones. So the setting up of the cone locations needed to satisfy the organization of spaces, views through the underbelly space, and the different span ratios of concrete and metal.[8]

In his essay "Arguing for Elegance," Schumacher promotes formal elegance as an articulation that achieves a reduction of visual complexity while preserving an underlying organizational complexity. For him, elegant compositions are highly integrable systems that cannot be easily decomposed into independent subsystems, much like natural systems where all forms are the result of the interaction of physical forces.[9] Design engineering similarly promotes an integrated process where shared knowledge from both disciplines is blended. The result is an interplay among forces in adaptation to engineering and architectural performances. This integrated design process creates an ambiguity in defining the work of each discipline. It is not the work of the engineer or the architect that determines the waffle slab pattern of Phaeno, but rather the collaborative effort that cannot be broken down into two parts.

Hybridization of Opportunities
When talking about the work of Rem Koolhaas, Sanford Kwinter compares his "extreme" architecture with a pilot flying a jet plane. The pilot, instead of being flesh and blood, is part of the mechanical realm of the plane. Only when the pilot is fully cognizant of the physical tolerances of the aircraft could this machine be maneuvered successfully, within physical limitations, in different directions.[10] The designer similarly grasps and utilizes the intuition of material continuity to find what is unseen as a source of novelty and creativity.

The analysis of the structure of Phaeno as a system was a big challenge for AKT that could be tackled only by using finite element analysis software. The software package (Sofistik), however, was not capable of analyzing the project because of its complexity and scale. This problem led to a crucial collaboration between the structural designers and the software engineers. The software package was updated continuously throughout the design process to cope with the complex load combinations. Without this tool, the structure would not have been able to meet the spatial objectives that the architects wanted to achieve. The software tool is an abstraction of the realities of structures on the construction site, and the design team didn't initially realize that even though the solved wall thicknesses from the computer analysis were enough to support the whole building, the walls didn't have enough space to vibrate concrete because of the density of the required rebars and the complex wall geometries.[11] This led the designers to use a nonconventional type of self-compacting concrete, often used for bridge and tunnel construction, that has the ability to flow into intricate formwork and pass through congested reinforcement.[12] Their understanding of the industry and knowledge of what was available in the market allowed the designers to apply this material to achieve the desired complexity for both architecture and structure.

As in the case of the Science Center project, to follow the "waves" of reality is to both direct and be directed by many factors. The constraints of matter are not obstacles to design but rather opportunities to discover what is unseen. Even though material reality seems to be rigid and constant, it is dynamic and continuously being updated with new compounds and hybrid opportunities, as in the examples of bringing an existing software tool to a more powerful level or applying a conventional material in a different composition.

The Wave of Information Technologies
The last two decades have created an ever-growing wave of information technology for designers. For engineers and architects, digital tools have become a standard part of their work processes for generating or evaluating design. When the first computer-aided software packages became commercial, the draftsperson was able to insure his or her work, as all the drawing information could be stored and replicated many times. This opportunity has led many practices to stop drafting with conventional methods. With the advent of finite element analysis tools, engineers gained the ability to analyze an increasing number of elements in a structural system while controlling for various parameters. Engineers and architects also use a number of tools for environmental and acoustical evaluation.

The digital space also facilitated collaboration among professions as they started to use the same medium of work production. Within the past thirty years, information-transfer technologies have evolved from fax machines to online platforms where sharing knowledge happens within seconds. Even though the tools of architects and engineers are different, software packages allow for smooth information exchange. Design engineers have developed their own interfaces to make the architect's model and their analysis model parallel, when standard embedded file-exchange tools are not sufficient. The more opportunistic engineers have been using the tools of architects, such as Rhino, and even making plug-in interfaces that allow them to see the architect's digital concept in its original informational form, to make a faster evaluation of it.

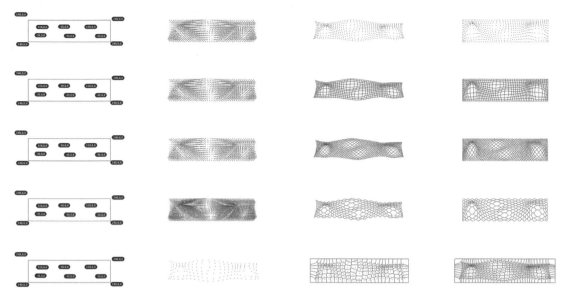

Computational Pattern Generation. Murat Mutlu, MIT master's thesis, 2010

There is also the concept of a meta-model, in which all information from the engineers and architects can be processed through one unified language. This collaborative platform has been achieved through software packages such as Catia or Digital Project. These tools also enable parameterization of geometric or scientific knowledge descriptions to be embedded as part of the meta-model, which allows the vision of the final product to become an element of the process, capable of incorporating complex sets of constraints.[13] This reciprocal relationship between product and process enables fast evaluation among multiple design alternatives. Like the standard CAD software insuring drafting work, parametric tools insure the amount of time that the designers invest in setting up the relationships between parts in a project. When the forces of reality push the designer to alter a design, whether due to a change in client demands or suggestions from engineers, parametrically related parts can be updated rapidly without reworking the old three-dimensional information.

The awareness of the power of information technology has also influenced academia. Student work from the MIT Computation and Design Group has been focusing on guiding this technology into a more malleable state for design flexibility, where design rules are laid out as codes of information. The designer's ability to conceptualize and translate the design of a building into geometric information is now taken to a higher level of abstraction, with the geometry defined as lines of script. This way of working enables a highly flexible and changeable design process—more than any standard CAD or parametric design package can facilitate. This technological momentum has affected the professional practices of both engineering and architecture firms as they started to house special design groups in their offices, such as the Advanced Geometry Unit at Arup, p.art at AKT, and the Computational Geometry Group at KPF, to name but a few.

The marriage between information technologies and sophisticated numerically controlled production equipment for making objects led to the development of production lines with high precision and the ability to materialize intricate forms while maintaining quality. Since the 1990s, there has been an explosion in the quality and quantity of computer-aided manufacturing technologies.[14] This technological evolution has helped architects with materializing their digital information to an extent where the product quality is determined by the quality and detail of the digital drawings. The ability to control the output of numerical machinery has become a form of contemporary craftsmanship, and this process of digital materiality has brought craft to design studios.[15] The lines of the architect can easily be transferred to the edges of the material with almost invisible tolerances; smooth digital surfaces can be milled out of blocks of material; robotic arms can produce consistent welded joints between materials.

The surface finishing on the concrete walls of the Phaeno Science Center is one such example where the final product was crafted digitally. The architects wanted to create a visual relationship between the cone geometries and the surface pattern created by the formwork.[16] The wooden planks of the formwork were drawn digitally, with tapered geometries, and then sent to CNC cutting machines to transfer the digital information to physical reality. The connections between the slabs and the cones were intended to be seamless; these doubly curved parts of the concrete geometry were achieved with a reinforced glass-fiber plastic molding method—again, a digitally informed process.

Designers have to be aware of the potential of digital tools so that they direct these evolving information technologies while avoiding their drawbacks. Information technology can propagate the process, but it does not generate ideas.[17] For the opportunistic designer, digital information processing is a test field for ideas in which the feedback between design and evaluation is much faster than without it, much as Cecil Balmond explains his process of design: "speculating an idea, probing and sketching it… do simple hand calculations and then I use the computer to prove the point."[18] The word "computing" refers to the procedure of calculating, of determining something by mathematical or logical methods. Therefore "computer" is a tool for processing logics that are defined by the user, but not a device for generating logics for processing. The logics and the constraints that need to be processed must come from the designer in the form of a speculated idea, probe, or sketch, in relation to real matter. The question of what the digital cannot perform needs to be critically considered in the design process. The contemporary design process simply cannot rely on the power of computational tools.

Conclusion

As knowledge in design and engineering continuously expands and the secrets of matter revealed, there is a growing necessity for information exchange between professionals with different expertise. This in fact is only possible through a collaboratively processed design where the constraints of materiality resolve into a matrix of solutions, which then become a blended proposition. The evolution of information technologies also has created a great platform for this integrated design process. Opportunistic designers, through any available means (information technology, computer-aided manufacturing, collaboration, etc.), use these constraints of materiality to spark innovation. Their abilities are not limited to any trend or style, as their design inspiration comes from the dynamic essence of reality.

Notes
1. Rem Koolhaas, "Elegy for the Vacant Lot," *S, M, L, XL* (New York: Monacelli Press, 1995), p. 937.
2. Alejandro Zaero-Polo, "Nontraditionality," *Design Engineering AKT*, edited by Hanif Kara (Barcelona: Actar, 2008), p. 60.
3. Lars Spuybroek, "Africa Comes First," *The Architecture of Continuity* (Rotterdam: V2, 2008), p. 125.
4. Cecil Balmond, *Informal* (Munich: Prestel, 2007).
5. American Institute of Aeronautics and Astronautics, *Aerospace Design Engineers Guide* (Reston, VA: AIAA, 2003), p. xi.
6. Hanif Kara, *Design Engineering AKT*, p. 1.
7. Patrik Schumacher, "Engineering Elegance," *Design Engineering AKT*, p. 74.
8. Notes from "In Search of Design through Engineering" course, lecture by Hanif Kara, Harvard University Graduate School of Design, Spring 2009.
9. Patrik Schumacher, "Arguing for Elegance," *Architectural Design* (January/February 2007).
10. Sanford Kwinter, "Flying the Bullet," in *Rem Koolhaas: Conversations with Students*, edited by Sanford Kwinter (New York: Princeton Architectural Press, 1996), p. 76.
11. Notes from "In Search of Design through Engineering" course, lecture by Hanif Kara.
12. M. Ouchi, S. Nakamura, T. Osterson, S.E. Hallberg, and M. Lwin "Applications of Self-Compacting Concrete in Japan, Europe, and the United States," ISHPC (available at www.fhwa.dot.gov/bridge/scc.pdf).
13. Kara, Design Engineering AKT, p. 9.
14. Dan Schodek, Martin Bechthold, Kimo Griggs, Kenneth Kao, and Marco Steinberg, *Digital Design and Manufacturing* (New York: Wiley, 2004), p. 24.
15. Spuybroek, "Experience, Tectonics, and Continuity," *The Architecture of Continuity*, p. 27.
16. Notes from "In Search of Design through Engineering" course, lecture by Christos Passas.
17. Simon Allford, "Re Design Engineering," *Design Engineering AKT*, p. 11.
18. Nina Rappaport, *Support and Resist: Structural Engineers and Design Innovation* (New York: Monacelli, 2007), p. 19.

Toward a New Sobriety: Rebel Engineering with a Cause

Evangelos Kotsioris

On October 28, at 9 am, I heard one of Balmond's engineers describe, without irony or noticeable wavering, how the encounter and eventual joining, at 200 meters, of sloping steel structures that, through their relative positions on the ground were exposed to different amounts of solar heat-gain, could only take place at dawn, when both had cooled off during the night and were most likely to share the same temperature. I was elated and horrified by the sheer outrage of the problem that we had set them. Why do they never say NO?"[1]

—Rem Koolhaas, on the construction of CCTV headquarters in Beijing

One of the biggest intellectual struggles in the creative architectural practice during recent years is probably that for a so-called new objectivity. It is debatable whether such an ambitious goal is realistic, especially when architectural decisions are increasingly tied to structural reasoning—the latter has come to be almost solely associated with notions of efficiency or a sense of an economy of means. It has become than obvious that the redefinition of the relationships among form, structure, tectonics, and ornament will be increasingly relevant but not easy.

Throughout the centuries, the relationship between architectural form and structure had always been ambiguous. Certain forms (arch, vault, dome) would indicate the limited possible ways that one could build them. Horizontal and vertical loads called for different systems, but their meeting points would always need to be smoothly coordinated. "Different building types were associated with different types of materials,"[2] which to paraphrase Sullivan, included all things organic and inorganic, all things physical and metaphysical, all things human and superhuman, all true manifestations of the mind, the heart, the soul.[3] Starting from the "ancient Greek concept of techné, understood as the rational basis for the construction of objects", and ending to "medieval ideas of the mechanical arts, which considered built forms as utilitarian objects".[4]

What is witnessed in architectural practice of roughly the last two decades is an unprecedented breach of this correspondence between form and structure. The reasons for this could be summarized broadly as the effects of great advance in computation and the globalized market economy.

Information technologies have radically changed the paradigm of all kinds of analysis, including structural. Apart from the fact that terms such as morphing, lofting, spline-networks, soft bodies, particle systems, force-fields, physics dynamics, and parametrics have entered the everyday architectural vocabulary, huge developments in simulation sparked an unprecedented tendency of substitution of every static behavior with a dynamic one. New areas of expertise exploded in highly specialized systems, resulting in "transport consultants, wind consultants, fire engineering consultants, access consultants, risk and opportunity managers, sustainability consultants, planning consultants, crime prevention design advisors, public relations and communication consultants, property advisors and letting agents, professional construction consultants, rights and light consultants, façade consultants—in addition to the more familiar ones such as structural engineers, mechanical engineers, cost

Rem Koolhaas/OMA, CCTV Headquarters, Beijing 2002-2011. Joining of the sloping steel structures

consultants, project managers, landscape engineers, construction managers, acoustic engineers, lighting consultants and so on."[5] The heroic era of the genius architect with an equally genius engineer devising a project by themselves is no longer plausible; today everybody wants to squeeze around the table, and everybody seems to have a good reason to do so.

At the same time, the return of shape in architecture came right on time to serve an unprecedented demand for iconicity favored by the market economy.[6] Koolhaas comments: "Junk-space is additive, layered and lightweight, not articulated in different parts, but subdivided [...] there are no walls, only partitions, shimmering membranes covered in mirror or gold."[7] Suddenly construction became simply too slow to serve the ever-changing demands of the market, which spawned an unprecedented amount of architectural production to serve its purpose, "depriving" architects of the luxury of slowness and reflection. The collage-istic approach that had already permeated the vertical "architectural" elements started to also sweep structure. Did "junk-space" start being staged around what one might call "junk-structure"? Koolhaas had already given a hint back in 2004: "Structure groans invisibly underneath decoration, or worse, has become ornamental; small shiny space frames support nominal loads, or huge beams deliver cyclopic burdens to unsuspecting destinations."[8]

Through the ill-defined process of "post-rationalization," almost every possible shape became buildable overnight. Traditional rules of thumb or simply "thicknesses" could fluctuate tremendously. It is no wonder that the fascination for production of architecture ("thick" by definition) through the use of surfaces ("thick-less" by definition) reached a climax during this period. The "craze" of the continuous folded surface of the 1990s (Rem Koolhaas/OMA's Educatorium and UN Studio's NRM Laboratories in Utrecht, MVRDV's Villa VPRO in Hilversum, or even Diller and Scoffidio's entry for the competition for the Eyebeam Institute in New York some years later) came with a whole new set of intellectual and tectonic riddles; to satisfy the basic principle of no differentiation between the-floor-that-turns-into-wall-that-turns-into-roof, architects and engineers would have to devise a series of "camouflage" techniques to mask the persisting differentiation between the structures needed for horizontal versus vertical loads. The construction of the first 50-centimeter-thick concrete walls to match the thickness of the horizontal floor planes made obvious the problematic nature of the folding tectonics, urging the "masters" of the technique to abandon it rather prematurely.

In the mid 1990s, the Bilbao Guggenheim can probably be considered the built apotheosis of the contemporary paradigm of complexity of structure and form. Analyzed to death—both in terms of structural engineering but also as the ultimate architectural/urban/economical phenomenon of the decade—Gehry's hyper-expressionistic shape seduced architects and the public. It had been a long time since a piece of "creative architectural practice" would spark such excitement in the public. This is probably Gehry's greatest lifetime achievement: intentionally or not, he managed to surpass the megalomaniac ambitions of all the traditional postmodernists combined.

The Bilbao Guggenheim pushed innovation in the use of digital design tools in architecture, but it also opened Pandora's box for structural engineers. What had been hyped at the time as "software used by the NASA aeronautic engineers" (Digital Project by Gehry Technologies, a simplified version of CATIA by Dassault Systèmes) was later appropriated by an increasing number of experimental architectural practices. It was through the new possibilities of this software that Bilbao managed to "transform" structure into a mere "holder" of shape put together by pinned or attached surfaces.

Frank O. Gehry & Partners, Jay Pritzker Pavilion,
Chicago 1999-2004

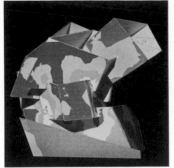

Studio Daniel Libeskind & ARUP,
V&A Spiral Extension, 2000. Finite
elements analysis model

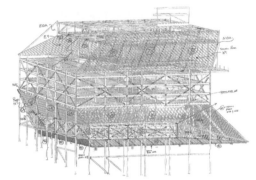

REX/OMA, Seattle Public Library, Seattle 1994-04

The unclad skeleton of the museum looked like a frame that had survived the brutal forces of a crash. This deformed frame, though, does not seem preoccupied with any kind of topological operation (as were those found in Eisenman's Cartesian deformations of the same period, for example) but rather acts as a suppressed, invisible carrier of ornament. Even though it would be naïve to plead for a new nostalgia for the "magnificent fight between weight and its contrary as we see it in the Parthenon", what's deeply frustrating in Gehry is the uncanny confrontation with, as Antoine Picon said, "a new kind of antigravity in the age of the superglue."[9]

It was at the end of the twentieth century though that the true implications of these changes in the conception and realization of structure and shape in architecture fully manifested themselves; the euphoric rainbow-colored diagrams of finite element analyses became hypnotic images on computer screens and magazine pages, as in the case of the V&A Spiral Extension by Daniel Libeskind, where Balmond argued that "structure and architecture become one immediacy."[10]

In the meantime OMA's blocky, programmatic foam models were translated into exoskeletal steel boxes, almost like some kind of irrational Constructivist structures. The excessive use of diagonal members needed to stabilize their unstable configurations, as in the case of the Seattle Public Library gave birth to the "aesthetics of bracing" (that will later find its apotheosis as the only projected elements on the outer skin of the CCTV Headquarters in Beijing). The crudeness of the geometry of the constructed parts and the complexity generated by their density

has little to do with the ruthlessly simple block models of the respective competitions, and that is rather desirable.

In the case of Casa da Música, Balmond and Koolhaas will go even further. Instead of using the "appropriate" thickness for the thin, load-bearing envelope, Balmond reduces it intentionally to create orchestrated "instabilities" that will be supported by a series of bizarrely protruding pillars growing out of the 1-meter-thick diaphragms of the main concert hall, reaching out to support precisely the points where the biggest stresses occur. The "peeled-out" model views of the project reveal the schizophrenic interior, where huge trusses are used to create the long-span roof of the auditorium, spiraling staircases unwind around the edges, bridges are supported by metal struts, and thin sheetrock creates temporary partitions between spaces. In a rare moment of such virtuosity, Koolhaas managed to combine all of the

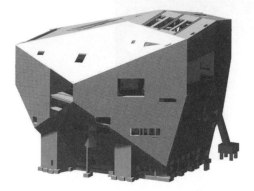

Herzog & de Meuron, Beijing National Stadium, Beijing 2003-08

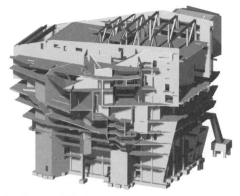

Rem Koolhaas/OMA, Casa da Música, Porto 2001-2005.
Renderings of the concrete envelope and internal structure

characteristics of "junk-space" and "junk-structure" in one building. Instead of trying to achieve the perfect "minimalist elegance," as Schumacher would say in a different context, he used all the insane apparatus made available to him to exorcise the junk-demons.

Almost in parallel, the Phaeno Science Center by Zaha Hadid Architects suffered from a similar schizophrenia of multiple systems (a main waffle slab, a Vierendeel roof supported by disproportionally "heavy" cones, prefabricated and cast in situ self-compacting concrete walls, metal columns on specific spots of "instability"), even though in this case the ultimate aim was the creation of an "elegance of complexity."[11] The reasons for such a broad range of systems apparently derived from implications of feasibility and cost.

The result is one of the most astonishing examples of architecture and engineering of the past decade, even though Hadid's ruthlessly internal architectural logic demands a holistic design approach in every single detail. This in turn leads unavoidably to a blatantly overdesigned project, where every external intervention appears as a punishable cacophony. The critical inquiry here could question the necessity for such complexity or even the efficiency of such design and analysis processes.

The most recent of all the phenomena that takes us back to the Bilbao Guggenheim would be the overtaking of the ornament. In the Beijing National Stadium by Herzog and de Meuron, the ornament is not just carried by the structure—it is the structure. The dimensions and optical characteristics of the ornament are intentionally designed to be practically indistinguishable from structure. As Koolhaas commented on the iconic project in "Venturian" terms: "It is the duck and the nest at the same time."[12] As entertaining as this may sound, the Beijing National Stadium is perhaps the first example in recent years where the fusion of ornament and structure challenge visual perception.

It seems as if we've reached a point where almost "everything" is constructible in a very short time, and architecture is still indulging this moment in time. As we move forward, the digital paradigm will only continue to alter our fundamental notion of tectonics. In this light, the question is not if the ornament will disappear anytime soon, but what will be the new paradigm in its coordination with structure.

Zaera-Polo argues that "A unitary theory of the building envelope may be an answer to the decoupling of politics and nature."[13] What is apparent is

Ali Rahim and Hina Jamelle/Contemporary
Architecture Practice, Commercial High Rise,
Dubai, 2007

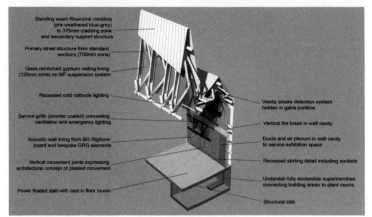

Zaha Hadid Architects, Glasgow Museum of Transport, 2004-10.
Axonometric section showing surface build-up

that the line between the envelope and its support will only become blurrier. According to Picon, it will become increasingly difficult to distinguish between the two, as new structures will get indeed rid of their armature in the traditional sense.[14] The "recent death of the diagram" is a hint of what we should expect. After a century of increasing the layers, is it maybe time to reduce them?

To create the new, more homogeneous and high-performance whole, one does not necessarily have to oversimplify and reduce these layers in number ("Less is More") because it is impossible to go back to the romantically simple detailing of Modernism. Instead, these layers could be reconfigured and developed to make use of their attributes in multiplicity ("More with Less"). We are confronted with the pressing need to find new ways of stratifying the layers of increasing complexity. Moussavi's notion of "expanded materiality" or "supermateriality" attempts to define the required "change in our approach away from an understanding of material as exclusively physical and tangible, to include both the physical and the non-physical; climate, sound or economics as well as wood, steel or glass."[15] In this paradigm, assemblages of structure could perform in multiple ways simultaneously.

Innovations in material technology can open up many possibilities, although what has been proven over time is that large-scale application will probably be one step behind. The ultra-high-efficiency materials we know, or new materials we expect to create, will still have to be composed or assembled in a way that creates synergies, drives performance, and facilitates large-scale implementation. We will have to continue in that direction before the new tools of each era are not merely mastered; where a "new sobriety" can be achieved and efforts are shifted from the surface to its actual depth. Instead of being a "rebel without a cause", an innovative attitude of experimentation with a serious purpose is needed, where a balance between energy spent and actual outcome is attempted. Where intelligent design engineering will work to reduce the maintenance of this increased complexity and seek the "elegance of smartness." Where multiple layers and surfaces actually collapse in one, one day.

Notes
1. Rem Koolhaas. "Post-modern engineering?", in Rem Koolhaas, Brendon McGetrick (eds.), Content (Köln: Taschen, 2004), pp. 514-515
2. Farshid Moussavi (et al.). The Function of Form (Barcelona; New York: Actar; Cambridge, Mass.: Harvard University, Graduate School of Design, 2009), p.7
3. Louis H Sullivan. "The Tall Building Artistically Considered", Lippincott's Magazine 57 (March 1896), pp. 403-9, cited in Kindergarten Charts and Other Writings (New York: Wittenborn Art Books, 1976).
4. Moussavi. The Function of Form, p. 9
5. Moussavi. The Function of Form, p. 7-8
6. R. E. Somol. "12 Reasons to Get Back into Shape", in Koolhaas Rem, Brendon McGetrick (eds.), Content (Köln: Taschen, 2004), pp. 86-87
7. Rem Koolhaas. "Junk-space", in Rem Koolhaas. Brendon McGetrick (eds.), Content (Köln: Taschen, 2004), pp. 162-171
8. Ibid.
9. Notes from course GSD 4355: Architecture, Science and Technology, XVIIIth Century-Present, Prof. Antoine Picon, Harvard GSD, Spring 2010
10. Cecil Balmond (et al.). Informal (Munich: Prestel, 2002), p. 191
11. Patrik Schumacher. "Arguing for Elegance", Elegance, AD (Architectural Design), January/February, 2007
12. "Flâneurs in Automobiles: A conversation between Peter Fischli, Rem Koolhaas and Hans Ulrich Obrist" in Hilar Stadler, Martino Stierli (eds.), Peter Fischli. Las Vegas Studio: images from the archives of Robert Venturi and Denise Scott Brown (Zürich: Scheidegger & Spiess, 2008), pp. 161-170
13. Alejandro Zaera-Polo. 'The Politics of the Envelope', Volume #17, Fall 2008
14. Antoine Picon. Digital Culture in Architecture (Basel; Boston: Birkhäuser, 2010)
15. Moussavi. The Function of Form, p. 8

Highcross

Leicester, UK

FOA / AKT

The John Lewis store and cinema at Hammerson's Highcross Quarter shopping center in Leicester was designed by Foreign Office Architects and engineered by Adams Kara Taylor. The project demonstrates how development practice and attitudes are changing in fostering collaboration, as well as highlighting design and structure relationships in mixed-use buildings with a common aesthetic identity and diverse programmatic requirements. The client selected FOA through a design competition, and the delivery of the winning idea demonstrates the value of such an approach while showing the process through design, contracts, and construction.

Facing several challenges in combining a mixed-use program at a site with archeological remnants, design and engineering collaborated to address diverse typological and spatial requirements for the different program elements. Primary entrances to the building are either through an amphitheater at one corner—where one reaches the theater multiplex—or by walking a 36-meter glass-and-steel footbridge that connects the shopping mall to a new multistory garage, spanning a six-lane ring road.

The interior experience is created by a combination of goods, sensations, and structural gymnastics. Several floors of seemingly structureless glass walkways rise through an atrium, and the corner itself stretches the potential of the reinforced concrete frame. The discussion on envelopes and patterns expands on the elaborate façade design that interweaves structural analysis with formal expressions of algorithms and materials. Two layers of panels of fritted glass and cut sheet metal come together to suggest lacework, reflecting Leicester's and John Lewis's shared history in haberdashery.

Ravensbourne College
London

FOA / AKT

Foreign Office Architects designed this new art college, just south of the Millennium Dome on Greenwich Peninsula, to recall the decoration tradition of the Arts and Crafts movement in a contemporary context. The idea is to create an abstract pattern from floral shapes that introduces a geometrical order based on a tile unit. This not only affects the façade but also determines the internal organization, such as floor-to-ceiling heights and structural grids. Behind this tiling is a perforated concrete wall that has considerable structural strength and thermal mass to assist with natural ventilation.

The scheme depends on creating an economy of scale by achieving repetition in the tiles across the eight-story building, with three basic shapes. The rest of the structure is relatively straightforward. Given the ground conditions, foundations are piled, while the precast concrete plank floors are supported on composite steel beams that span 15 meters from the perimeter walls to a central spine of columns to create longer column-free spaces.

The college provides studio spaces as well as the scope to open voids for volumes that need more height than the standard floor-to-ceiling space, such as an atrium, a television studio, and a lecture theater.

GSD Lite

Francisco Izquierdo, Giorgi Khamaladze,
Jarrad Morgan, and Stephanie Morrison

Weight versus Lightness

Harvard University as an intellectual and cultural institution carries significant weight. The name alone conjures images of solidity and permanence. But is this the correct image for the Harvard Graduate School of Design, an exhibition center for contemporary architecture intended to embody notions of openness, accessibility, and agility? We propose to balance the prevailing impression of weight, epitomized by the heavy architecture of The GSD's Gund Hall, with an exhibition space that is an expression of lightness, drawing inspiration from the canopies of trees that occupy the site. Lifted into the canopy and playing with light and transparency, the center for exhibiting contemporary architecture will reflect a current attitude to design practices that values openness, transparency, and responsiveness—lightness itself.

Harvard versus Cambridge

Established in 1636, Harvard University is the oldest institution of higher learning in the United States. The school has continually expanded over the past 400 years throughout Cambridge, Massachusetts, and, not surprisingly, this has caused occasional tension between the institution and its host community. Conflict arose in 2008 over Harvard's proposed expansion in Allston, just across the river from Cambridge, a project opposed by Allston residents. Although the University and the city hold differing opinions on many issues—zoning, planning, garbage control, taxes, environmental impact—the town-gown tensions in Cambridge are similar to those in other cities.

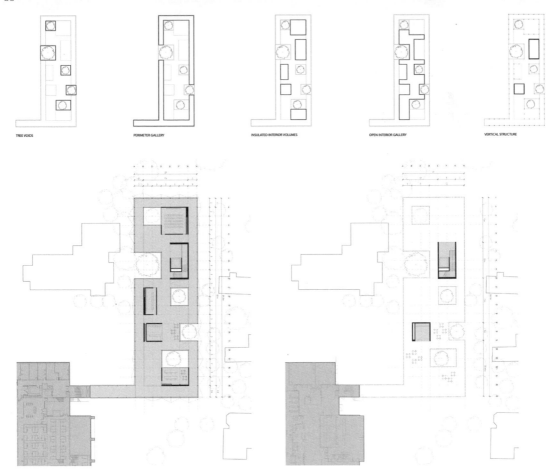

TREE VOIDS   PERIMETER GALLERY   INSULATED INTERIOR VOLUMES   OPEN INTERIOR GALLERY   VERTICAL STRUCTURE

Institutional versus Residential Fabric

Harvard's academic and laboratory facilities are of a much greater scale than the surrounding residential fabric (much of this fabric is also part of Harvard's real estate portfolio). There exists a fine and tenuous balance in Cambridge between these two scales. To respectfully maintain the finer-grain residential scale (and good graces of the community), we will eliminate only one building to site the new GSD exhibition space. The building will sit toward the rear of Gund Hall, sharing its back lawn, and face Kirkland Street, where the public entries to the galleries will be located. At ground level, passage through the grounds will be both accessible and highly public. In response to the varied climate of Boston, we propose a twofold strategy to address environmental performance.

Thermal Cross-Section - Summer and Winter

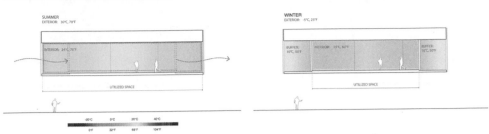

Adaptive Program - Summer and Winter Mode

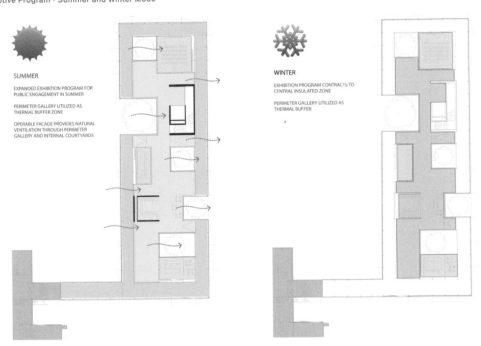

Adaptive Programming

Closely tied to the operation of the Harvard Graduate School of Design, the new Center for Contemporary Architecture will be subject to fluctuating use that reflects the academic cycle, yet the cycle of public engagement will follow a different course. During the semester, GSD Lite will serve primarily as an extension of GSD programs, offering a place for exhibitions, displays, and debates. In the summer, when Cambridge receives an influx of tourists, we propose to expand the center's exhibition and gallery program.

Wrapping the perimeter of the elevated plan, a linear gallery will be activated, capitalizing on the warmer weather as a shaded, naturally ventilated transitional space—a verandah—providing some thermal buffer between the exterior and the more controlled microclimate of the interior spaces. During the winter, these programs contract and pull back to the core spaces in the center of the plan which are more heavily insulated and controlled, leaving the perimeter gallery to act as a thermal "buffer zone" providing an intermediate micro-climate between the heated interior and the cold Boston winter outside.

Perforated/Folded Aluminum Facade Screen

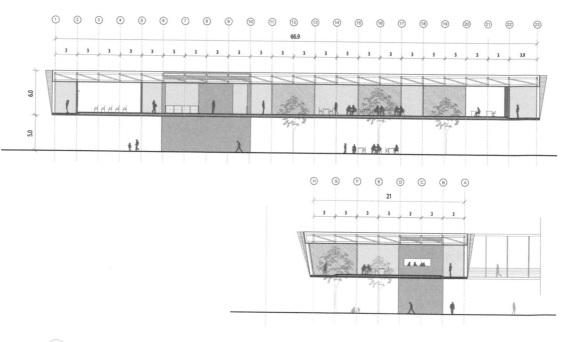

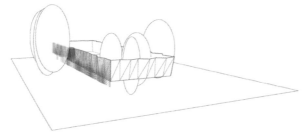

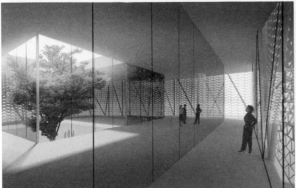

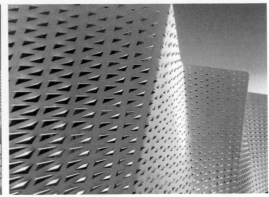

Performative Skin

Lifted up into the canopies and accommodating the existing within glassed glazed courtyards, the project takes advantage of the natural shading. Yet the structure's with aspirations toward transparency, care must be taken to limit solar heat gains through the extensive areas of glazing. A highly responsive perforated screen to the exterior is proposed, changing in density in response to the need for protection from the summer sun. Orientation and inclination have been used to determine the size and orientation of the perforations parametrically, as well as the location of the existing trees and their shade. The perforations thus evolve along the length of the façade, reducing in size in areas of exposure and opening up to light and views in areas afforded protection by the tree canopies.

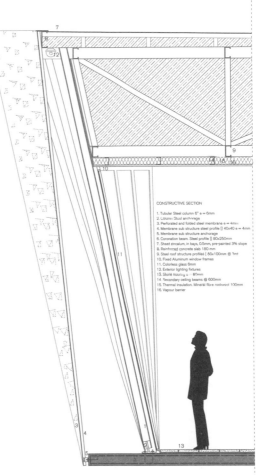

CONSTRUCTIVE SECTION

1. Tubular Steel column 6" e = 6mm
2. Column Dtuel anchorage
3. Perforated and folded steel membrane e = 4mm
4. Membrane sub structure steel profile [] 40x40 e = 4mm
5. Membrane sub structure anchorage
6. Coronation beam. Steel profile [] 80x250mm
7. Sheet zincalum, in bays, 0.5mm, pre-painted 3% slope
8. Reinforced concrete slab 180 mm
9. Steel roof structure profiles [] 80x100mm @ 1mt
10. Fixed Aluminum window frames
11. Colorless glass 6mm
12. Exterior lighting fixtures
13. Stone flooring e = 40mm
14. Secondary ceiling beams @ 600mm
15. Thermal insulation. Mineral fibre rockwool 100mm
16. Vapour barrier

Detail Section

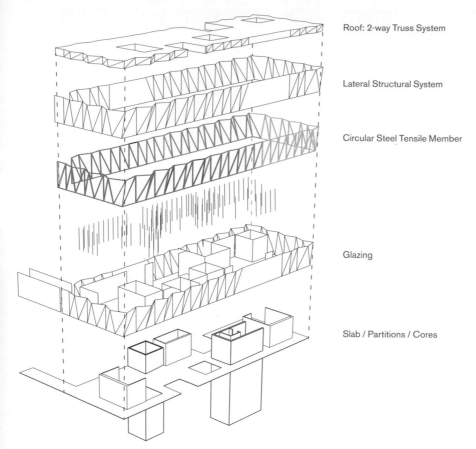

Roof: 2-way Truss System

Lateral Structural System

Circular Steel Tensile Member

Glazing

Slab / Partitions / Cores

Structural Strategy: Strength from Above

To achieve a sense of lightness, we propose a structure whereby the elevated floor containing the majority of the program is suspended from above. With just two contacts with the ground in the form of concrete cores, the raised floor appears to hover, almost impossibly disconnected from the ground. While they provide points of entry and activation of the ground plane, the two concrete cores also support a trussed roof, set on a 3 meter x 3 meter grid, which in turn suspends the elevated floor set 5 meters above the ground through a grid of steel members. As a system exploiting the tensile strength of steel, these vertical supports for the elevated floor can be kept to minimal cross-section dimension, adding to the impression of lightness. To combat potential issues relating to lateral wind loads in areas farther away from the cores, the vertical supports on the perimeter begin to shift in the diagonal to provide lateral bracing, subsequently creating ripples in the alignment of the façade. The large cantilevers at either end of the building, 15 meters and 18 meters to the north and south respectively, are made possible by the thick roof truss, concealed to a large degree by the perforated screen. The elevation of the structure and subsequent liberation of the ground plane provide Harvard and the community with an extended public park that addresses Kirkland Street as a continuation of the existing GSD lawn.

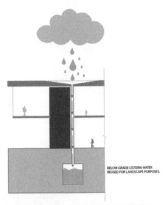

RAINWATER HARVESTING SYSTEM: DOWNSPOUTS LOCATED ALONG
WALL OF CORES.

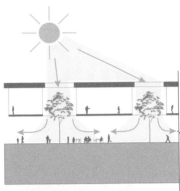

TREE VOIDS CARRY ADDED BENEFIT OF DAYLIGHTING UNDERSIDE
OF BUILDING AT GROUND LEVEL, INCREASING ITS APPEAL AS A
SPACE FOR PUBLIC ENGAGEMENT.

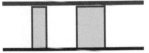

CONCRETE CORES TO ACCEPT
ROOF STRUCTURE

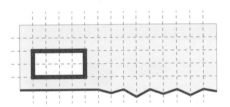

FAÇADE SCREEN & SUPPORTS: PLANAR WHEN ADJACENT TO CONCRETE
CORES; INCLINED WHEN AWAY FROM CORE TO BRACE AGAINST LATERAL
LOADS

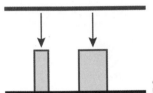

ROOF STRUCTURE RESTS ON
MASSIVE CONCRETE CORES

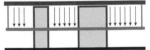

FLOOR HANGS IN TENSION FROM
ROOF ASSEMBLY ABOVE

SUSPENSION MEMBERS: VERTICAL ADJACENT TO CORE / INCLINED WHEN AWAY FROM CORE

Harvard M.O.D.E.

Travis Bost and Werner Van Vuuren

THE MODULAR OPENING FOR DESIGN AMD ENGINEERING

A platform & pavillion for communities and international expos

HISTORY

> Largest art collection, limited exhibition space
> Thriving overlapping community, no social spaces
> Expansion, and implosion, in Allston
> 3 Communities. The campus core and periphery
> Cyclical / Seasonal, Nature of Harvard
> Harvard's history with tents.

Design precedents

Re-imagined Chauhaus

Versatility & Design

GSD / HAM Anchor

The proposed building functions as a platform and pavilion. It capitalizes on Harvard University's vast art collection and provides a solution to the limited exhibition space of the institution.

By creating a space for exhibitions and community gatherings, the project brings together the thriving art scene of Cambridge and Allston. By addressing three distinct communities—the student campus, the Cambridge neighborhood and the surrounding areas—the project can have several cyclical or seasonal exhibitions, matching the life of an academic community in the New England climate.

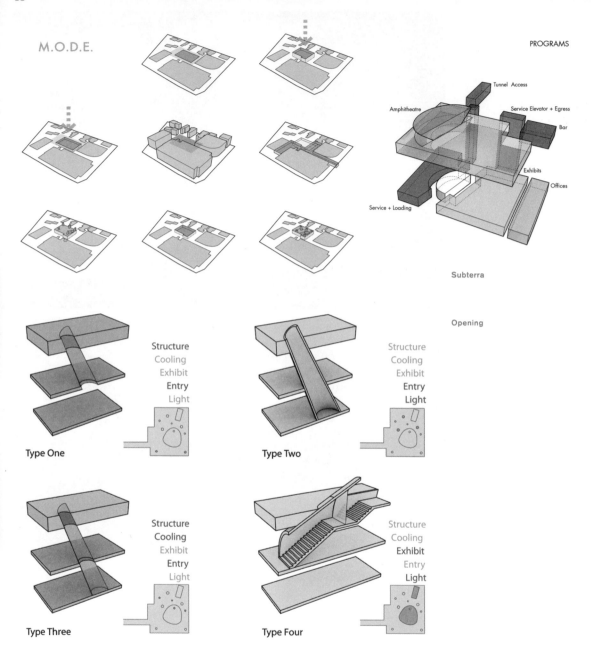

M.O.D.E.

PROGRAMS

Tunnel Access

Amphitheatre

Service Elevator + Egress

Bar

Exhibits

Offices

Service + Loading

Subterra

Opening

Structure
Cooling
Exhibit
Entry
Light

Type One

Structure
Cooling
Exhibit
Entry
Light

Type Two

Structure
Cooling
Exhibit
Entry
Light

Type Three

Structure
Cooling
Exhibit
Entry
Light

Type Four

The programmatic goal for this project was to minimize the impact on the limited available communal open space of this area of Harvard's properties that encompass a number of diverse departments. Pairing this need with the limited necessary daylighting requirements of gallery space, it was beneficial to place all program below ground.

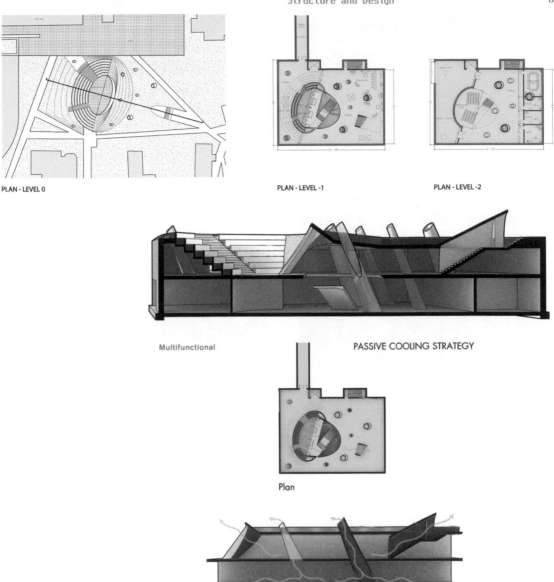

PLAN - LEVEL 0

PLAN - LEVEL -1

PLAN - LEVEL -2

Multifunctional

PASSIVE COOLING STRATEGY

Plan

Idealized Section

Similarly, the out-of-scale imposition of the program-specified bridge to link back to the Graduate School of Design was rethought as a less obtrusive underground tunnel (though still subordinate to the primary ground-level entrances). The project therefore became a manipulation of a handful of formal and operational vocabularies that negotiated the building's public dialogue and that of its internal function and loads. Thus the entrances become pronounced public areas that let in daylight, act as an informal gathering amphitheater space, and host a seasonal pavilion.

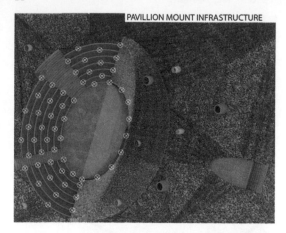

PAVILLION MOUNT INFRASTRUCTURE

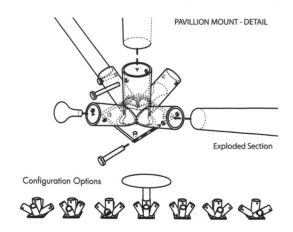

PAVILLION MOUNT - DETAIL

Exploded Section

Configuration Options

Versatility

Similarly, three types of light and air wells puncture the gallery space, each with at least two functions. Some privilege the exchange of fresh air or heating/cooling needs, others deliver light to the most recessed spaces, and still others act as exhibition spaces themselves; additionally their emergence from the turf at ground level is a curious folly to the users above, which some lightwells play up by becoming bench seating, while those of glass are seen as curious windows into the exhibits below. Finally, these many forms of lightwell act together in a structural capacity with the two-way slab floors that are able to span the area columnlessly given the lightwells' effective function in replacing them—an ideal situation for flexible gallery space.

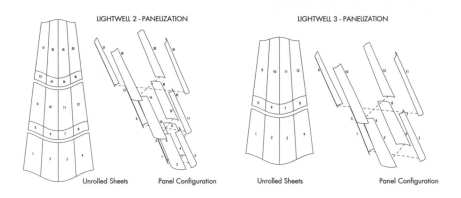

LIGHTWELL 2 - PANELIZATION

Unrolled Sheets   Panel Configuration

LIGHTWELL 3 - PANELIZATION

Unrolled Sheets   Panel Configuration

Implementation

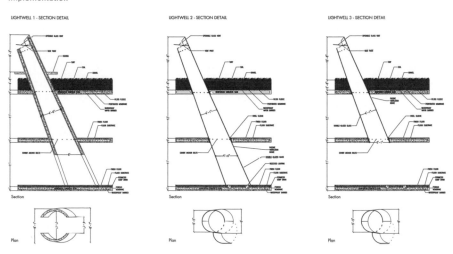

LIGHTWELL 1 - SECTION DETAIL

Section

Plan

LIGHTWELL 2 - SECTION DETAIL

Section

Plan

LIGHTWELL 3 - SECTION DETAIL

Section

Plan

TERRACE - SECTION DETAIL

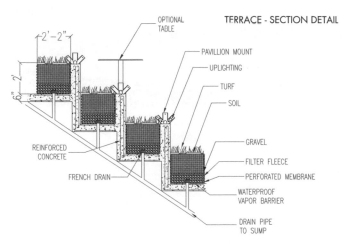

OPTIONAL TABLE

PAVILLION MOUNT

UPLIGHTING

TURF

SOIL

GRAVEL

FILTER FLEECE

PERFORATED MEMBRANE

WATERPROOF VAPOR BARRIER

DRAIN PIPE TO SUMP

REINFORCED CONCRETE

FRENCH DRAIN

2'-2"

2'

6"

Phaeno Science Center
Wolfsburg, Germany

Zaha Hadid Architects
AKT

Design the Cloud: Cross-disciplinary Approach to Design

Christos Passas

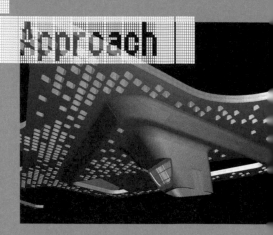

Discussions on the ground of intersection in disciplines related to Architecture or Engineering give us the possibility to think outside the box and to consider how traditional disciplinary boundaries are formed; the notion of each discipline and inevitably the everyday constraints of our professional practices. We theorize on the potential benefits of multiple intersections and the merits that potentially exist.

These intersections already exist in many forms despite the judicial and business separations that are imposed by the legal system. However, the production of work on complex projects, require today a more integrated work method based on communication and information exchange; a design protocol that blends the boundaries of the stereotypical parallel disciplinary approach.

I have to admit that my initial response to this writing was that there was no need for justification of a cross-disciplinary method of working. Zaha Hadid Architects and Adams Kara Taylor have worked successfully on the Phaeno Science Centre—and other projects since—and these projects, verify the case of this method of collaboration. In designing complex projects, where an intensified cross-disciplinary approach is a necessity, the intersection, between practices is important. However, there are several problems, the most important of which is that each discipline uses different thinking criteria and values in their Design approach.

In this sense, I welcomed the opportunity to contribute to this book, and discuss the merits and problems of developing a cross-disciplinary collaboration discourse.

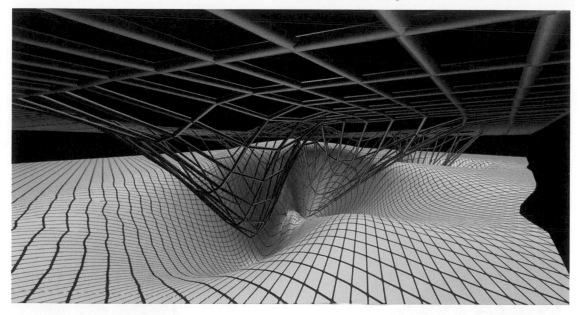

Imagination to Reality

Despite its obvious bi-polar, Descartian origins, I have to admit that I was always fascinated by the notion that a work of art or design, from its inception to its realization is a transition between two worlds: From the virtual (sphere of pure thought) to the real (sphere of the corporeal).

One way, in which this transition works, is by exchanging approximation (loose, fluid entities) for precision (stable, fixed entities) during the design process. This exchange is in essence the focusing of the concept to a particular manifestation which is enriched by the integration of multiple parameters, such as systematization, rationalization, materiality, material economy etc. While these latter mechanisms can exist in the sphere of the virtual as pure strategies, they are more effective in their ability to guide design concepts and transform them into real world manifestations. In that way they are effective devices for design development in that they can describe precisely the geometrical and material characteristics of the object of design.

Research and Experimentation

The role of pure Research is crucial to Design. Technology and experimentation is a prerequisite to Innovation. However, one must also acknowledge that Innovation can be achieved both as product of "specialist" thinking and as a product of pure Necessity. Practice can be a very fruitful field for the development of such on-site research. Similarly, experimentation is crucial for developing knowledge integration skills, alternative design thinking methods and for testing innovative ideas and the plausibility of concepts.

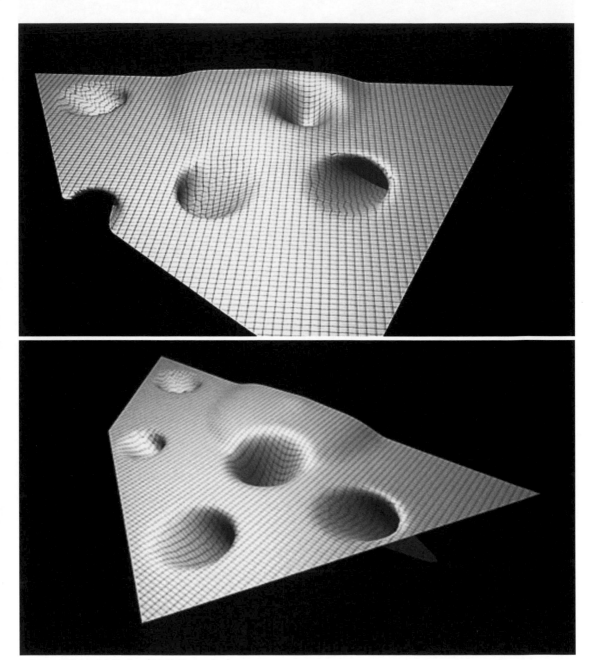

Instances of Concept Animation; (dot. Matrix + surface)
deformations. Phaeno Science Center

Design Tools

Historically, the work of forming the building and the work of structuring it were one and the same.

The distinction in disciplines simply did not exist until the industrial revolution. Architects and Engineers, since, engage in practice, using classical ways of working. Each discipline maintains each respective profession's impenetrability, using predominantly 2d tools for the representation of their thinking. Le Corbusier's and the Domino house prototype effectively worked within the paradigm of the rational separation of the disciplines. Structure could be separated from the form of Space and Space would be freed from Structure.

The working methods emerging in the beginnings of the 21st century give opportunities for relooking at this model and speak of an alternative: A model of multi-disciplinary collaboration in which design is based on networked information economies. The digital tools that are extensively used today are partly responsible for the way work is made but the tools don't make the work. Collectively speaking, this fundamental shift in design thinking is also the reason for designing the tools in the first place. Therefore today, we are effaced with a thinking crisis and not a dilemma of how to use the digital tools.

The shift to 3d tools and the introduction of computer technology enables designers not only to share information more quickly, with higher degrees of precision but also enables a different way of thinking of Form.

Performance, integration, and learning to navigate interconnectivity are crucial to the education of designers today. The emergence of systems that can at once perform on multiple threads of information within a highly visual environment provides designers with an invaluable new ways of imagining and delivering projects.
How can this information handled to deliver results?

Working Methods

While the working methods are not necessarily interesting in themselves, it is important to distinguish between the two most prominent: The Classical and the Adaptive Approach. The Classical approach focuses on the fragmentation of a pre-analysed (theorized) problem and the intelligent interlinking of specialists focused on solving aspects of a particular project. It works its way through the project, in a rigorous manner, effectively by breaking it down into its constituent parts and resolves it by delivering a set of solutions for each part and then sums them up in a definitive whole. The Adaptive approach does things a little differently. By maintaining openness to extraneous information the particular design network (say the design team) absorbs and incorporates external information into its workings. It remains unresolved, longer, as to what the final solution might be, but works in a mode of continuous integration towards a desired end. Each mode, delivers in a different and exclusive way. The classical (specialized) approach can lead to deep research and deep design solutions, while the adaptive (generalist) approach allows for a more fluent, absorptive performance in dealing with high levels of abstract information while classifying them into of packets of informa-

Instances of Concept Animation;
(dot. Matrix + surface) deformations.
Force fields applied on multiple surfaces

tion that feed into the design object. The adaptive approach can have multiple loops in the process. Overall, both approaches are linear in the way the work is executed.

A third approach can be the employment of a mode of Flickering from one approach to the other: A design thinking approach that can constantly employ imagination and creativity to integrate and manage the flow of information while working on the refinement of the project. In this mode, the designer is not a disciplinarian but a catalyst that can think inside and outside the box in a perpetual shift between the two modes described above. Mark Dziersk[1] explores the notion of Design by comparing its common meanings as an end result and as an action. He describes design thinking as "…A protocol for solving problems and discovering new opportunities. Techniques and tools differ and their effectiveness is arguable but the core of the process stays the same." He also goes further to define a process of design thinking that consists of a four steps: 1. Define the problem, 2. Create and consider many options, 3. Refine selected directions (and loop) and 4. Execute

I would go a step further to say that in the design of buildings and urban projects the process of design takes in many cases a few years to be complete. It is therefore impossible to go through this cycle only once. While the process of refinement can run in parallel, my experience tells me that the process of design follows throughout that time frame constantly imagining new solutions and integrating the project.

Information Management and the Design Process
We can witness the operation of interdisciplinary processes in the world today when we consider the working methods of organizations like NASA, ESA and other large scale institutions.

It is well known that in these organizations deliver very complex projects. One is initially, left in wonder as to how thousands of scientists, designers, engineers, thinkers, problem solvers operate effectively. What structure might enable them to operate in their tasks? In order to achieve the degree of innovation and forward thinking that is necessary, Information must be organized in multiple layers and hierarchies in order to perform the multiplicity of the particular tasks.

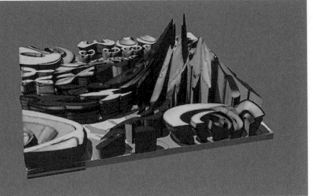

The project of launching an international space station that is developed to operate in orbit for long periods of time requires an unimaginable degree of organization and coordination. With information coming in from all sides of the scientific spectrum, disciplines as remote as biology, computer science, particle physics, engineering, psychology etc the participate in the task of integration. Information management goes part of the way to deliver the required results.

In no way, is there an attempt here, to establish of a simplified relation between information and knowledge. The new information economies are by nature both in flux and interlinked and this gives rise to an interesting opportunity about the way in which information "terminals" handle information flow, within the design network.

Information flows continuously, in an endlessly faster pace, delivering a sensation of ever-changing sensibilities, so designers, must maintain a degree of openness. This openness not only allows for continuous update, but can inform the design development and the regeneration of knowledge already owned. A becoming notion is the emergent nature of knowledge.

The Emergent Nature of (Design) Knowledge
The emergent nature of design knowledge has to do how the system collectively works. First, it is important to establish what the designer's operation is, in connection to the form making process. The designer inevitably places themselves at the intersection of receiving external information (brief, restrictions etc) and making a series of internal value judgements in regard to the design. This simple action can be repeated and at once form and inform both sets of values (external and internal).

Collectively, these decisions are shared, either as packets of code, as step-by-step design processes or as "design concepts" that are proliferated to the design community. This simple process then becomes an emergent system. Once a design solution gets accepted by the community of designers as a viable solution to a problem, it gets distributed in a bottom-up way and forms a "trend", i.e. an acceptable way, or a performance protocol for small or large solutions, within the community of designers.

Development of a Common Language

In order to consider how to enlarge the intersection between disciplines, it is important to develop a language or a protocol of compatibility. It is evident that different disciplines think about problems in different ways. Immersing design disciplines in a common environment may be a productive way to reduce the differences without losing out on specialization. Which environments are most conducive to producing integrated designs of high level complexity within a multidisciplinary design approach? Is it a virtual environment, a physical environment or is it a hybrid condition? How can environment be used to enable conceptualization and growth of such design cultures? How can a design process that effectively manages a large amount of problems, in some cases contradictory, be organized in order to bring disparate thinking patterns together? Should there be a predominant thinking structure or one that (e)merges (from) the input of different specialists? Is it possible to think of structures that communicate and collaborate, whilst

thinking of problems in different ways? How can we achieve high level integration and latent complexity or better sophisticated simplicity? Geometry is certainly one tool which designers use to communicate ideas precisely and effectively and this must be the starting point. Furthermore, certain geometrical concepts are more adept towards promoting this culture in that they necessitate intensive collaborations.

Examples/ Building
The Phaeno Science Centre is an example of such a concept, where the architectural Form, developed as a Play between Surface and a system of Forces that caused it to deform can immediately be translated into a Structural concept.

These deformations play out as an exploration into the single surface project. This particular experiment is in a way not a three-dimensional form but it is conceived as a purely four dimensional object, i.e. an animated entity. Therefore, the surface deformations are not figurative as they don't form a static composition but effectively become the parameters of the design. The conditions produced by the forces (deformations) become the tool with which the first instances of the design manifest.

These deformations are spread wide forming a complex network of conditions (generalist/ approximation) that are elastic enough to be able to maintain integrity even under the most scrutinised level of rationalization. In effect, the project becomes a case of synthesizing the effects of "forces" into a complete result. The "force field" of conditions becomes a simultaneous translation into spatial formations, structure, occupation density and so forth.

Surface flexing for instance, may be rationalized through the design development process and in the Phaeno particularly interpreted into the conical volumes expressed as an interior crater landscape that essentially creates space and structure together. The concrete surface simultaneously spread wide and composed itself into a form that becomes the image of this force field.

Therefore, the Design is only evident as a manifestation of the Form and it becomes both an architectonic and a structural device. Generated as abstract means at first, they become instruments for a transformation from the virtual to the real.

In this way, the Design engagement is one that allows parametric design processes to be effectively productive, away with traditional composition methods or the figurative portraiture of static object observer compositions and effectively working on a malleable set of conditions. The object is no longer made the way it is seen, but the way it is thought or felt.

This apparent Strangeness is evidence of its authenticity as an experiment and in the best of case an emergent form of beauty.

Examples / City:
In the case of Sofia Masterplan, the experimentation is taken a step further to incorporate a larger set of elements in a unified whole with distinct building elements that are based on algorithmic functions.

By nature a Masterplan is amongst other things, an assembly of buildings. These building and their interrelations have to be organized and correlated, void public space must be generated and the effects of the urban conditions assessed in a simulated environment. Furthermore, the environmental credentials of the urban plan must be integrally developed and modelled both in terms of appearance and performance.

In this case, the intermittent spaces are generated by the stretching a tentative urban pattern. The effects of this elasticizing the urban mass can generate buildings that are no longer based on a zoning diagramme. The transition from concept to refinement and execution can be implemented in a lateral way or through looping sequences.

This method allowed us to develop the Masterplan and the building plan simultaneously, while other issues can immediately follow such as the analysis of façade elements not for individual but for an array of buildings, prior to the final formation of the programmatic contents themselves or indeed the overall urban form.

This simultaneous development of elements can, for example, test the performance of façade surfaces, (see figures 7-8), prior to the finalization of the building form, thus enabling the designers to instantly assess issues such as solar gain, shading conditions, viewing possibilities and transparency while forming the urban plan.

In some sense, the design of the project could become a live feed model by bringing in other aspects to study such as the phasing of construction, the calculation of investment volume per phase, traffic engineering or the exploration of shared infrastructure concepts in quantifying the effects of each design decision.

Conclusions

Cross-disciplinary design may not only be a tangible possibility but also a necessity in our working methods today. The evolution of design thinking from the parallel multi-disciplinary process to a unified integrated thinking process would include fluid information exchange, deep interrelations between disciplines and design interaction that continues from the beginning to the end of the project.

This level of collaborative design thinking can certainly prove itself worthy in experiments using CAD/CAM technologies and rapid prototyping innovations that are currently being developed. The future will most certainly hold the demand to assemble buildings remotely by delivering virtually designed models of buildings directly on site. The information based collaborations and the introduction of very powerful computing tools in the design process offers a possibility of evolving design practice from a craft, or a design studio to a building laboratory.

The issue now is, not only how to tap into this valuable resource of merging technology and our ability to manage large volumes of information, but how to elevate our imagination and deepen our thinking to correspond to these high demands and produce beautiful highly integrated designs.

Notes
1. Article "Design thinking…What is that?" published in www.fastcompany.com March 20th, 2006

Design systems allocate resources in response to various factors and contextual constraints. Within a building, several apparent or implicit functions can coexist, with each system responding to a different set of variables. Optimal fit to context can be an inherently difficult design process when several systems overlay and vie for control of limited resources such as space, circulation, program, void, structure, energy, and light.

Contextual and Systemic Design

Changing Forms, Changing Processes

Dimitris Papanikolaou

Is it the form that drives the design and production process, or the processes themselves that determine forms? While often design expression pushes engineering ingenuity to invent new solutions, it is typically technological innovation that offers the tools to designers to explore new formalistic domains. Through the course of history, design practice has been integrating technology, people, and materials to invent new methods to increase form customization while decreasing production costs. We are witnessing a transformation within the building industry of what was previously known as empirical craftsmanship to today's highly controlled digital fabrication. Nevertheless, today's digital technology seems to raise as many questions in design practice and research as it answers.

The Industrial Revolution

The industrial revolution established standardization, mass production, and prefabrication in the building industry. Large machines effectively replaced manual labor in simple, repetitive tasks, while skilled workers concentrated on the cognitively complex tasks of assembling, handling, and distributing. Spurred by the machines' high setup costs, industries standardized components, processes, and shipping methods while production volume increased dramatically to benefit from economies of scale. Prefabrication factories used more sophisticated production and logistical processes to remotely fabricate, preassemble, and deliver building parts to the construction site for final assembly and installation. Panelized modules with universal interfaces could be combined in multiple ways like Lego blocks and flat-packed inside shipping containers to decrease project delivery time and supply-chain costs. Suppliers and fabricators specialized in products and services, formulating collaborative alliance clusters in geographically larger market networks.

Mass production and prefabrication systems created highly centralized supply chains, while their cost efficiency depended heavily on location and shipping volume. Design representation methods focused on Euclidean orthographic projections of floor plans, cross-sections and elevations to share technical information between contractors and designers, spreadsheets to order materials and estimate costs from suppliers, and perspective drawings or physical models to communicate ideas to clients. Furthermore, increasing standardization and repetition of building forms disengaged designers from studying custom structural details, as most technical specifications were now predefined by the industrial suppliers and building contractors, who could often change the design outcome significantly.

The Digital Revolution

The digital revolution has seamlessly integrated design and manufacturing, allowing designers to build a 3D CAD model on a computer and directly fabricate it using a CNC machine. Initially developed for the aerospace, naval, and automotive industries, digital design and fabrication are now transforming the building industry, reshaping processes, forms, and services. Widespread availability and decreasing cost of personal computers, fabrication machines, and software packages is making digital design and fabrication increasingly accessible to designers, engineers, contractors, material suppliers, and building product manufacturers.

Digital Design

Computer-aided design (CAD) replaced hand drafting by allowing designers to automatically create, modify, and reproduce digital drawings on demand. The first CAD programs had a non-associative modeling approach, such that modifying a component of the model had no impact on the rest of the geometry. This made the design process rather tedious and time-consuming, as the designer had to manually adjust each component of the model. Modern CAD programs changed the designing process from a mere geometric representation of unassociated forms to a functional description of processes that can generate those forms. This shift from representation of forms to description of processes is a fundamental concept in modern computational design practice, as computer programs now follow the design instructions and remodel the resulting geometry for different input parameters. Modern computational design thus eliminates redundancy and opens the doors of complex geometric modeling to designers.

Computational modeling methods today can use either generative or parametric design approaches; often, however, a CAD model will combine both of them. Generative design focuses on algorithmically creating geometric forms by programming a list of instructions in a scripting language that, once executed by a CAD compiler, produces the resulting forms. The scripting language manipulates primitive geometric components such as points, lines, planes, and surfaces using variables, functions, conditional statements, loops, and grammar rules. Generative design deploys a bottom-up approach that can produce topologically different results, an aspect that makes it popular in form-finding and form-optimization techniques. Parametric design, on the other hand, is based on hierarchically associating geometric components of the model with mathematical equations such that modifying any input parameter of a component propagates changes in the entire model. Parametric modeling has a top-down approach, moving from the "parent" components (often called the driving geometry) to the "child" components, while their topological structure remains unchanged.

Often an entire parametric model can be encapsulated and used as a component in another parametric model.

A computational design model can thus be considered a black box that takes input arguments and outputs a resulting geometry. Such black-box models can be linked to external databases to store and exchange properties that in turn can be further linked to other models, programs, or collaborators, creating dynamic workflow chains that automatically update, each time a change in one of the links occurs. Designers thus can model complex geometric forms and instantly change the thickness of the walls, the profile curve of a beam, or the density of a structural grid, updating in real time the entire model and the exported properties lists, without endless hours of rework.

Computational modeling, however, is also an experience-based skill. In contrast to traditional non-associative geometric modeling, there is no single approach to building a computational model: the same result can be reached by different parametric or scripting approaches, but the level of control will be different in each case. Selecting the appropriate design method to build a computational model depends on thoroughly understanding the design requirements and available data; availability of a clear project scope and contextual parameters can distinguish a good modeling strategy from a bad one.

Digital Fabrication

Digital fabrication or computer-aided manufacturing (CAM) uses computers to digitally control high-precision fabrication machines (CNC) to build physical prototypes from CAD files. A special computer program translates the input CAD file into a tool path, while a control system in the machine drives the motors of the tool tip along the path during fabrication. Digital fabrication methods can be either subtractive or additive, depending on whether they remove material from a monolithic block (milling, laser-cutting, plasma-cutting, etc.), or instead deposit material into stratified layers (3D printing, fused deposition modeling, etc.).

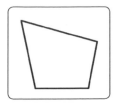    

Digital fabrication production workflow

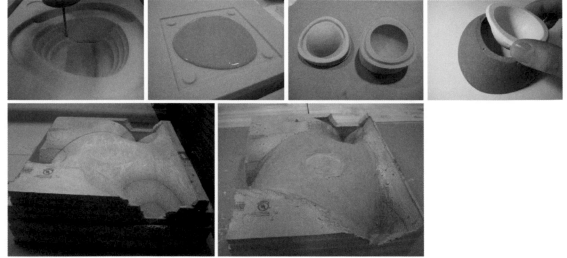

CNC milled molds and formwork can be used to cast plastics or composite materials

Depending on the number of axes and motors that move the tool tip, a fabrication machine can have two, three, four, or five degrees of freedom, with a significant impact on fabrication capabilities. Typically the tool tip is moving relative to a stable bed, the dimensions of which limit the maximum size of a part that can be machined. However, research today uses autonomous mobile robots equipped with CNC fabrication machines and sophisticated geo-positioning systems that can navigate in space fabricate similar to the drawing turtles of the classic LOGO language.[1]

Computers, digital fabrication machines, and assembly stations can be combined to create very efficient production systems. Workflow starts from the development of a master model, rationalization and decomposition of the master model into detailed part files, distribution of the part files to the fabrication units for machining, shipping of the finished parts to the construction site, and final

assembly. These processing steps can be linked through programs, computers, machines, and humans, creating a dynamic production system that can make almost anything, anywhere, and at any time at a cost depending mainly on material, equipment type, machining time, and shipping distance.

The digital fabrication supply chain is spatially and functionally decentralized. As local fabrication facilities and internet communication means spread around the world, a CAD model being developed in North America can be electronically sent to fabrication shops in Asia for prototyping and shipped to a nearby construction site for assembly, significantly decreasing transportation costs. Furthermore, as digital fabrication machines become smaller, smarter, and cheaper, an entire fabrication unit can fit inside a shipping container and be sent directly to a construction site, further lowering shipping and rework costs.

Phaeno Museum in Wolfsburg: CNC milled formwork

Guggenheim museum in Bilbao: Complex geometric forms decomposed into developable surfaces

New Forms

The digital revolution has had a tremendous impact on both building forms and design strategies, giving designers new perspectives, but also creating new construction challenges. From the doubly curved titanium-clad forms of the Frank Gehry's Guggenheim Museum in Bilbao to the interlocking plywood panels of the Instant House of MIT in the Museum of Modern Art in New York, the typical digitally fabricated building is a geometrically complex assembly of both customized and standardized parts. Generally speaking, the greater the number of standardized parts, the less flexible the design, but the easier the construction due to repetition and economies of scale; on the other hand, the greater the number of customized parts, the more flexible the design but the harder the construction. Based on this trade-off, a typical challenge in digital fabrication is to determine the line between standardized and customized parts in a geometrically complex design: at one extreme, complexity is uniformly distributed in the parts, such that each part is slightly different from the other; at the other extreme, complexity is strategically concentrated in few highly customized parts, while the rest of the parts are standardized and repetitive.

Another design challenge in construction of free-form geometries is to ensure that all complex surfaces are decomposed into smaller developable surface panels that can be easily fabricated through cutting, bending, and forging flat sheet materials. This means that surfaces cannot have double curvature; their unidirectional curvatures should not exceed the maximum bending curvature

DRL pavilion in London: Complex assembly of interlocking planar parts

that can be achieved by the available materials and techniques; the angles of polygonal panels should not be overly acute; and surface continuity between individual panels must often be retained to provide a smooth finish.

Another compelling design constraint is to ensure that converging planar ribs at either the nodes or other boundaries of a free-form structural grid will retain to the greatest extent possible their topological and angular relationships as the overall driving geometry of the model is modified. This is necessary to simplify and standardize joint and fastener detailing as well as to transfer loads smoothly between parts. For example, often the planar ribs of a quadrilateral structural grid must be locally perpendicular to both the skin surface they support and the other ribs that converge in the nodes. The list of constraints in computational design for complex geometries can be long, depending on the requirements of each project, making digital design and fabrication a highly specialized and challenging field.

The Stitchyak: a digitally fabricated kayak using a stitching technique to facilitate assembly

Design for Manufacturing and Design for Assembly
Seamless integration of design with production has led designers to consider how the latter could benefit from the former. Emerging from the fields of industrial design and product development, design for manufacturing (DfM) and design for assembly (DfA) are strategies that utilize design intelligence as a means of facilitating fabrication and assembly processes. For example, the studs (the small raised portions of Lego bricks) allow a child to easily position and assemble a series of Legos even with closed eyes. An excessive increase of the number of contact points between two parts, however, can make positioning and assembly difficult and time consuming. Similarly, using snap-fit joints, self-locating tabs and grooves, and registration marks facilitates assembly while minimizing the need for required joints, fixtures, and formwork. DfM and DfA engage designers in thinking of the assembly process as an integrated part of the design, a rather forgotten art since the advent of industrialization.

The following examples illustrate cases in digital design and fabrication where processes, tools, and materials inspired new forms and strategies. *Stitchyak* is a digitally fabricated kayak created a the MIT Media Lab utilizing a stitching technique with temporary zip-ties that helped with positioning and alignment of the curved cut sheets during assembly, without any fixtures or formwork. The stitched sheets were afterward covered by fiberglass for waterproofing and structural durability, and the zip-ties were removed. Such a process would otherwise have taken considerable time and

The Fabcar: a digitally fabricated 4WD toy car whose complex mechanical assembly consists of 234 unique parts

effort, without taking into account the construction of the formwork itself. Stitchyak was parametrically modeled, allowing for customization in form, size, and curvature to fit different body types and styles. Special consideration was given during the design process to the curvature and grain direction of the plywood surfaces to guarantee that they would be easily bent without cracking.

Fabcar is a digitally fabricated four-wheel-drive toy car made at the MIT Media Lab entirely out of plexiglas sheets shaped with a laser cutter and manually assembled without adhesives or fasteners. Fabcar's complex mechanical assembly consists of 234 unique parts organized into three differential gear mechanisms to unevenly distribute torque applied from a top central shaft to each of the four wheels based on their relative torque resistances. The design of Fabcar used snap-fit flexure joints that bend to ensure easy installation and spring back to trap the installed parts and

The Instant House

The built Instant House at MoMA's back yard

prevent accidental disassembly. Material stiffness and tolerance were empirically estimated to take into account the material removed during the machining process.

In the same spirit, the MIT Digital Fabrication Group designed and built the Instant House, a digitally fabricated house that was exhibited at the Museum of Modern Art in New York as part of the "Home Delivery" exhibition in 2008. The Instant House was a proof of concept of a new design and construction system that uses a generative shape grammar to create customized connection details for each plywood panel based on its location within the overall house geometry, and a portable three-axis CNC mill that can be transported in a shipping container to the construction site to fabricate the panels. The Instant House was made entirely of plywood sheets using glue-less, friction-fit, notch-and-groove connections; assembly was done on site by two nonskilled workers in a few days. The structural system of the envelope consisted of a dense quadrilateral grid firmly connected to an external and internal sheeting layer. Instant House had one main design constraint: since three-axis milling machines can only cut perpendicularly to the surface of a sheet, the structurally connected parts must be either perpendicular to each other or coplanar. As a consequence, since the planes of the structural ribs should always be perpendicular to the internal and external sheeting layers, they should also be perpendicular to the envelope's edges, where neighboring wall faces meet. Furthermore, the structural grid should be continuous for smooth load transfer between parts. Although successful in principle, the Instant House

generative system had several limitations: there was no firm geometrical solution in the CAD model to parametrically adjust the structural grid's layout as the envelope's geometry changed; the assembly hierarchy was not organized into subassemblies, a fact that made the assembly process tedious and difficult.

Further studies at MIT developed YourEnvelope, a generalized computational solution to this problem that parametrically readjusts the structural grid configuration as the envelope geometry changes ensuring that the perpendicularity constraints are always met. YourEnvelope could take as input the angles of the faces, the desired grid density, and the thickness of the material and output the cut sheet files with customized connection details. Furthermore, the overall structural grid was cleverly organized into nested subassemblies such that the assembly process could be conducted more quickly and easily.

New challenges
Despite the positive impact on design and production, the digital revolution also brought new challenges to construction. Not surprisingly, many digital fabrication projects have been construction nightmares, either due to logistical mismanagement of the numerous tasks and practitioners or assembly incompatibilities at the construction site. Many fabrication projects take more time than originally planned, are more expensive than expected, involve great risk and uncertainty, and

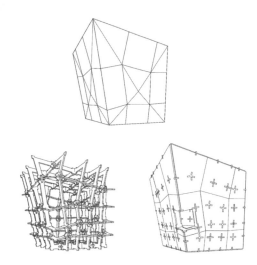

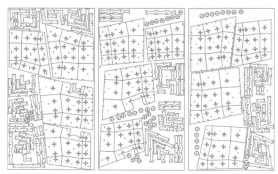

Cut sheet layout of the 252 parts of YourEnvelope

YourEnvelope: Parametrically adjustable structural envelope consisting of 252 custom parts organized into: (a) structural grid consisting of 12 subassemblies of 8 parts each and 48 connecting joints; (b) skin paneling consisting of 108 panels (54 interior and 54 exterior)

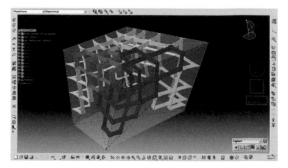

Parametric variations of YourEnvelope's geometry and grid organized into different subassemblies

prove to be too complex to plan, understand, and manage. Moreover, most problems are discovered at the construction site, when it is already late for corrective actions.

Evaluating the constructability of design can be a hard task, requiring skills and tools that we are just beginning to explore. It depends on the installation vectors of the parts during assembling and also on the number of connections between different parts. In practice, assemblies are studied through CAD modeling and physical mockups. 3D CAD modeling represents the final state of the assembly, however, when all parts have been put together, but not the process of putting these parts together. The order of constraint delivery in parametric CAD models is not necessarily the same as the constraint delivery of the actual assembly. As a consequence, by studying a 3D CAD model, the designer cannot easily tell if a design is constructible or estimate the difficulty of the assembly sequence. Physical mockups are typically used during design development to test constructability, but with a significant loss in time and cost.

Testing is empirical, understanding the solution to the geometrical problem is obscure, and design development becomes intuitive. While considerable research and technology have been invested in digital design and fabrication, empiricism and intuition characterize assembly at the construction site. Studying assembly is a matter of analyzing the topology of the assembly graph of the connected components. The following example illustrates a case where the liaison graph, which shows the order of constraint delivery between the parts, was used as a means to understand assembly incompatibilities.

Fabseat was a project done at MIT in 2007 to explore constructability assessment problems. Fabseat is a digitally fabricated and manually assembled chair consisting of 29 interlocking plywood profiles that together formulate a doubly curved structural surface. Fabseat was

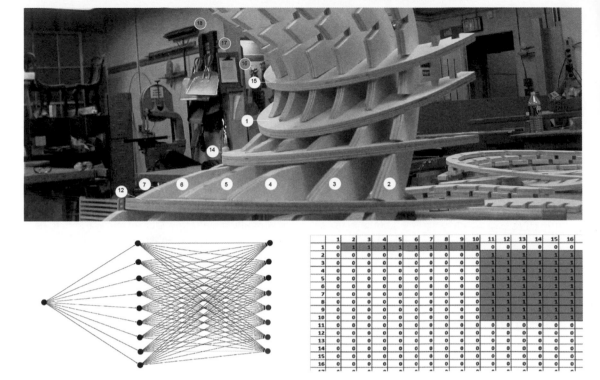

Assembly sequence analysis of the Fabseat highlighting the parts that could not be installed

	1	2	3	4	5	6	7	8	9	10	11	12	13	14	15	16
1	0	1	1	1	1	1	1	1	1	1	0	0	0	0	0	0
2	0	0	0	0	0	0	0	0	0	0	1	1	1	1	1	1
3	0	0	0	0	0	0	0	0	0	0	1	1	1	1	1	1
4	0	0	0	0	0	0	0	0	0	0	1	1	1	1	1	1
5	0	0	0	0	0	0	0	0	0	0	1	1	1	1	1	1
6	0	0	0	0	0	0	0	0	0	0	1	1	1	1	1	1
7	0	0	0	0	0	0	0	0	0	0	1	1	1	1	1	1
8	0	0	0	0	0	0	0	0	0	0	1	1	1	1	1	1
9	0	0	0	0	0	0	0	0	0	0	1	1	1	1	1	1
10	0	0	0	0	0	0	0	0	0	0	1	1	1	1	1	1
11	0	0	0	0	0	0	0	0	0	0	0	0	0	0	0	0
12	0	0	0	0	0	0	0	0	0	0	0	0	0	0	0	0
13	0	0	0	0	0	0	0	0	0	0	0	0	0	0	0	0
14	0	0	0	0	0	0	0	0	0	0	0	0	0	0	0	0
15	0	0	0	0	0	0	0	0	0	0	0	0	0	0	0	0
16	0	0	0	0	0	0	0	0	0	0	0	0	0	0	0	0

designed using conventional non-associative 3D modeling software. During assembly, about a quarter of the parts could not be installed without warping the material, an issue that was impossible to detect during 3D modeling development. A network analysis of the assembly sequence using the liaison graph that was done afterward showed that some of the nodes were impossible to install because they required more than two simultaneous installation vectors that were not parallel.

New Practices

Digital design and fabrication have significantly affected professional practices, as the designers of complex geometric assemblies must holistically take into account the machine, material, and computational constraints during design process. The complexity of design and construction of digitally fabricated buildings creates a new type of professional specialist who combines educational knowledge from design, computational geometry, programming, manufacturing, and structural engineering, among others fields. Many design firms around the world have formulated special interdisciplinary groups of experts who work as the interfaces between designers and industry to bring the most demanding projects to life.

Other emerging professional practices are specializing as external consultants on building information modeling (BIM), offering technical expertise and project management to traditional offices that lack this knowledge. This emerging type of specialist creates an unprecedented need for changes in the educational system, as the multidisciplinary nature of the field does not fit within any existing engineering or design discipline. Furthermore, digital technologies in design have affected contractual relationships between architects, engineers, and digital design specialists, as determining liabilities and intellectual property rights in cross-collaboration platforms such as BIM is not always straightforward.

From Digital Fabrication to Digital Materials
Our digital design and fabrication capabilities are already able to materialize the most complex geometric forms that our modeling skills and imaginations can create. As we expand the scale and scope of our fabrication technology and decrease its cost, we will soon be able to equip those forms with intelligence for sensing, thinking, and reacting to better respond to changing environmental needs.

Today new additive digital fabrication methods can print composites with variable structural and chemical properties by fusing materials with different properties. New digital electronics fabrication methods can place circuits with microcontrollers, sensors, and actuators into building components that can sense applied loads through force or vibration and realign their molecular fiber structure to better withstand those loads via electric signals. Moreover, smart building components can talk to each other and propagate messages through the entire assembly, turning potential buildings into habitable distributed computing platforms. Our future goal should be to apply these emerging technologies in meaningful ways in architecture.

Notes
1. www.hexapodrobot.com; http://el.media.mit.edu/logo-foundation/logo/turtle.html

Project acknowledgments: The Instant House research project was done by Professor Lawrence Sass and other members of the Digital Fabrication group of MIT. Stitchyak, Fabcar, and YourEnvelope projects were done by Dimitris Papanikolaou at the MIT Media Lab. Fabseat design and assembly sequence analysis was done by Dimitris Papanikolaou, while fabrication was collaboratively done by Josh Lobel and Magdalena Pantazi at the MIT Design and Computation group.

Bibliography
Anderson, David M. *Design for Manufacturability and Concurrent Engineering* (Cambria: CIM Press, 2006).
Bergdoll, Barry, and Christensen, Peter. *Home Delivery: Fabricating the Modern Dwelling* (Basel: Birkhäuser, 2008).
Bechthold, Martin; Griggs, Kimo; Kao, Martin; Schodek, Daniel, and Steinberg, Marco. *Digital Design and Manufacturing: CAD/CAM Applications in Architecture and Design* (Newark: John Wiley and Sons, 2005).
Huang, Jeffrey. "Interorganizational Information Systems in Design." Doctor of Design thesis. Graduate School of Design, Harvard University, 1997.
Mitchell, William J., and McCullough, Malcolm. *Digital Design Media* (New York: John Wiley and Sons, 1995).
Ohno, Taiichi. *Toyota Production System: Beyond Large-Scale Production* (New York: Productivity Press, 1988).
Sass, Lawrence; Michaud, Dennis, and Cardoso, Daniel. "Materializing a Design with Plywood." ECAADE, Frankfurt, Germany, September 2007.
Shingo, Shigeo. *A Study of the Toyota Production System* (New York: Productivity Press, 1989).
Whitney, Daniel E. *Mechanical Assemblies: Their Design, Manufacture, and Role in Product Development* (New York: Oxford University Press, 2004).

Enhancing Prefabrication

Sylvia Feng

Prefabrication and custom design have traditionally been viewed as opposing ends of a spectrum –a choice between efficiency and design freedom. Architecture is considered a personalized product tailored to the specific needs of the user. Mass-produced architecture is considered to be an "off the shelf" packaged solution that is designed not just for one client, but for an average of client needs. When one examines the trajectory of new technologies emerging in the field of mass production and industrial design, however the distinction between custom and factory-built is eroding. A vast number of tools available to architects inform a multifaceted, layered method of design by allowing the designer to think along many different lines at once. Buildings can be at once custom solutions, a sculpture, a piece of industrial design, a constructible form, and an expression of engineering.

In the early days of mass production, the key to the efficiency revolution was designing a series of small, easy tasks that shaped and assembled raw material incrementally into the final product. The process of making an industrial product was broken down into simple tasks for laborers with little training or skill to perform at their stations on the production line. Machines along the assembly line were designed to perform repetitive measured tasks such as lifting, stamping, and mixing. This type of production is most efficient in making many similar or identical products. The most illustrative example of the efficiency of mass production in architecture are mail-order houses in America. Sears Roebuck sold more than 70,000 pre-designed catalog homes in the early 1900s by bundling all necessary building materials and shipping the package to the client's plot of land. The house had already been designed, the materials were measured and cut to size, and all necessary fasteners were included. The remaining task on site was assembly, which required less expertise and therefore cut down on construction time and cost.

The next iteration of the catalog home is the "cookie-cutter" development made well known by Levitt & Sons' developments, commonly called "Levittowns." Instead of shipping homes to the owners, vast swaths of land were converted to neat suburban communities. While Sears Roebuck featured hundreds of designs, Levitt & Sons offered six models, making mass production of these homes even more efficient. The site essentially became the factory; the building process was broken down into simple tasks and identical houses replicated. The exterior and interior finishes provided options for customization by the owner. At this moment, when mass production merged with architecture, architects started to take on the role of industrial designers; their houses were also products. The traditional direct business exchange between owner and designer is managed by a middleman. The owner of the house does not have any interaction with the designer, at all and customization of the architecture is confined to a catalog of design options.

It is clear that the more assembling done within the factory setting, the easier it is to control quality and cost. Motor homes and trailer homes, dating back to the 1920s, are examples of architectural products completely assembled in a factory. Designed to be moved to location via highway, these architectural products dealt with size and weight constraints that were never issues in traditional architecture. Industrial architectural hybrids such as the Airstream motor home married engineering and design to generate a unique and influential product.

Aerial view of Levittown, PA

Airstream

Habitat 67 by Moshe Safdie

Mass-producing volumes that can aggregate into buildings has been a subject of experimentation and innovation. Habitat 67 by Moshe Safdie is a prominent case study showing the potential as well as the shortcomings of aggregated preassembled volumes. The original program called for 1,000 residential units on the Saint Lawrence River in Montreal, Canada. The units were aggregated in a staggered manner to give each unit a terrace. The units were self-supporting in terms of plumbing, mechanical, and structure and were tied together with steel cables. This is a case where the choice to prefabricate shortchanged the full realization of the project. The source of many problems was the decision to use concrete, a fluid building material. It is difficult to control, as its precise qualities depend on the consistency of the mixture in each batch as well as the environmental conditions in which it cures. The design itself did not allow enough tolerance to absorb the inconsistencies between pours, making it difficult to join units. The concrete units were also extremely heavy, requiring a lot of crane time to assemble the building. Transporting the units was not possible due to their dimensions of the units and a factory was set up on site to produce the units. What was intended as an economical construction method resulted in units that needed to be priced higher than most high-end real estate due to the unforeseen construction difficulties. This project was meant to be affordable housing, but the city of Montreal could not afford to complete the complex. Instead of 1,000 units, 158 apartments were constructed using 354 prefabricated units.

The complex eventually became privately owned and converted to condominiums. The adaptability of the units was limited by the structural capacity of the volumes. Since the volumes supported themselves, options for breaking through the concrete walls are limited. Despite the struggles along the way, Moshe Safdie accepted and worked within the many constraints that come with prefabricated units to produce an evocative piece of architecture that also pushed the development of mechanized building.

In the 1970s, Kisho Kurokawa and several other Japanese architects founded the Metabolist movement. The thesis behind the movement is that buildings can be designed with the flexibility to adapt and update, much like living evolving creatures. The Nakagin Tower (completed in 1972) is fourteen stories high and used 140 self-contained capsules. The prefabricated units were small enough to be transportable by highway, then bolted to the main shaft in the center of the building, using just four bolts. The ease of installation of the units as well as the choice of using bolts instead of welding was intentional to allow new units to be swapped into the building. The central supporting spine was built on site, while the capsules were manufactured off site at a factory. The capsules were made of steel members clad with steel panels. Tolerances were easier to control through the use of standardized materials that can be milled to precise dimensions. The building was originally meant to be an economical hotel for businessmen, and the capsules arrived complete with cabinetry and furniture installed. The units have yet to be "unplugged" as Kurokawa proposed; decades after it was built, however, the Nakagin Tower still holds great potential to adapt to new programs and even take on a new aesthetic identity.

Nakagin tower building exterior   Original Nakagin tower hotel
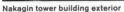room interior

Contour Crafting concrete printed
corrugated wall

There is clearly still an interest in housing as product today. Modern examples of prefabricated units for housing usually come in either flat "kit-of-parts" form (where pieces can be put together to form one unit), or in volumetric form. The kit-of-parts versions lend themselves to for low-to midrise structures. They aggregate easily in a planar fashion and are easy to assemble on site. Companies such as FlatPak and Ikea (under the name BloKlok) offer such kits. Since each "module" is one flat component that is wall, ceiling, or floor, the modules are easily customizable in terms of finish and materiality, thus creating mass customized products. Prefabricated pieces are made indoors at a factory, so construction and progress does not rely on outdoor site conditions.

Quality control can be managed with construction details such as spacing between layers and tolerances. Services are integrated at the factory by either designing components with slots for piping to slide into once the unit is set in place or including the piping within the component, for connection to service mains. Another advantage is the ease of transport. As opposed to finished units, flat components can be easily shipped, though they require more on site labor to assemble the components.

The kit-of-parts approach to prefabricated buildings allow for considerable freedom in cosmetics and cladding as well as expandability. Plans can be extended infinitely in two dimensions. The aesthetic and structural flexibility of the building is constrained, however by using a fixed number of component types. To make component kits cost-effective, varia-

tions among pieces are kept to a bare minimum, not allowing for structural variation to handle irregular loads or heavier loads, for example, if one wanted to build a mid-to highrise complex using the kit. The layout of the building must first follow structural constraints. The partitions used for the lower stories should be stronger than the panels used on the upper stories. Wall components are built to be load-bearing walls, but can aggregate only a couple of stories before extra bracing is required. Shear walls must be arranged in a way that loads are transferred efficiently down to the foundation; the plans are therefore restricted to arrangements that allow shear walls to stack linearly or cross at the same intersection at every level.

In recent years, mass-production technology has rapidly changed in tandem with our understanding of computing. Drafting and modeling programs are not only for documentation but also for experimentation. Digital modeling programs and fabrication tools have become extremely sophisticated, especially in the industrial design and engineering field. A new evolution in architecture is pushing past the confines of mass-production machinery. Architects are exploring parametric programs (such as CATIA or Grasshopper), prototyping machinery (three-dimensional printers and scanners), and CNC routing capable of cutting out custom components large enough for construction.

Programmable machines are able to translate increasingly complex commands and perform more sophisticated tasks in series. The biggest leap in

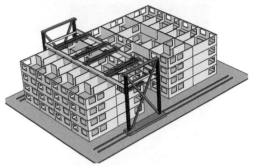

Contour Crafting illustration of 3D printed building complex

development is the ability to vary what the machines can produce without sacrificing efficiency. CNC routing allows for customized, individual components to be accurately produced in the same amount of time it would take to make several copies of a single component. It is possible to cut a range of materials used for architectural applications including woods, aluminum, plastics, composites, and foam. In Industrialized and Automated Building Systems, the author explains the capabilities and range of robotic machines and how they can be coded to do anything from mixing and dispensing concrete to brick laying and even drilling in fasteners.[1]

These production machines are becoming reality. For example, research currently conducted in Loughborough University in partnership with Foster and Partners in the United Kingdom is developing a concrete printer. With the input of a digital model, full-scale components could be printed in mortar using material dispenser nozzle moving on a track. Similar technology is being developed by Behrokh Khoshnevis of Contour Crafting, who imagines this technology being engineered into full-scale printers capable of printing entire residential structures one after another. These technologies not only hold potential for sparking innovation in design but can also efficiently provide mass housing in times of need.

Automated building cuts down on material waste since error is reduced and cuts can be arranged to minimize scraps. By thinking of architecture in terms of components and customized large-scale products, one can build in features commonly found in the industrial design field, such as the ability to upgrade or interchange. Closing the life-cycle loop of construction materials has been ignored in the field of architecture and results in the production of for 20–25 percent of landfill waste.[2]

According to the Environmental Protection Agency, construction waste includes concrete, wood, asphalt, metals, bricks, glass, plastics, and excavated trees from site clearing. Materials are thrown away during construction and buildings are demolished using methods that leave no chance for harvesting reusable materials. The attention that is given to the assembly of a building can also be performed in reverse. Products as simple as a flashlight and as complex as an airplane are designed to be assembled and also disassembled and refurbished. By considering the end of a building's life cycle as much as the beginning, materials can be designed to be disassembled from components in a form intact enough to be reused or to be disposed of properly. Structures almost always outlast the users or function of the building. The owner or operator of the building can change, or there may be a need for a different program. By designing flexible and interchangeable architecture, the life of a building can be extended. By rethinking the way we construct, architecture as a discipline can evolve.

Considering the plethora of production and design tools available to designers today, we are now able to reconsider not only the forms we can create but how rich in purpose architectural work can be. Collaborating with other disciplines is easier as architects, through their tools, can incorporate more types of data during the design process. The old method of factory production is now more than the ability to make architecture in quality; it is an experimental platform where we can learn how to make better architecture. Instead of being limiting, our tools can finally be enabling.

Notes
1. A. Warsszawski, *Industrialized and Automated Building Systems* (New York: Routledge, 1999).
2. climatex.org

Bibliography
Bender, R. *A Crack in the Rear-View Mirror: A View of Industrialized Building* (New York: Van Nostrand Reinhold Company, 1973).
"Habitat: Some Lessons." *The Canadian Architect*. October 1967.
International Council for Building Research, Studies, and Documentation. *Towards Industrialized Buildings* (Amsterdam: Elsevier, 1965).
Safdie, Moshe. *For Everyone a Garden* (Cambridge, MA: Massachusetts Institute of Technology, 1974).

Design for Disassembly: Closing the Materials Loop without Sacrificing Form

Scott Silverstein

No matter how well a structure is built, it will not last forever. Structural engineers long ago developed the idea of a "service life," in which a structure is designed to be structurally durable for a given number of years after construction (commonly fifty years for buildings today).[1] But modern buildings rarely last even that long—owners change, demands change, and there inevitably comes a day when a structure no longer serves its original purpose. Moreover, end of life for a building generally means end of life for the bulk of its constituent materials. Conventional construction methods create heavily integrated building systems that cannot easily be dismantled piece by piece, so most buildings are demolished, reduced to rubble, and sent to a landfill.

The standard life cycle of a building is a linear process (from construction to useful life to demolition), and eliminating waste and reducing demand for new materials requires converting this into a cyclical process that maximizes reuse and minimizes waste of resources. Design for Disassembly (DfD), also known as Design for Adaptability and Deconstruction, is an emerging initiative that reflects this goal. In particular, DfD believes that what happens to a structure at the end of its life should be considered in design. Given that a new structure will one day be torn down, designers have a responsibility to plan the building such that waste is minimized at the end of the building's life.

Three facets of DfD will be addressed here. First is adaptability: a building's spaces should be easy to reconfigure for different uses, extending a building's service life indefinitely. Second is deconstructability: it should be physically and financially possible to take a building apart and recover its constituent components. Third is reusability: the salvaged components of a building should be easy to use again, requiring minimal additional manufacturing. Notable among sustainable construction initiatives, DfD is particularly concerned with structural systems, which comprise more than 50 percent of the building weight and, traditionally, have the lowest potential for reuse.[2] DfD stresses an increased collaboration of the architect with the structural engineer, as well as the service engineers (responsible for designing building utilities, HVAC, and other systems), to make reuse a viable option.

Extending the Service Life

The term "service life" usually refers to the period for which a building's structural systems are designed to carry their loads: a technical upper limit on the building's useful existence. At the end of its technical life, a building is usually unsafe for habitation, and its structural components may have deteriorated such that they cannot be salvaged. But designers, as well as real estate owners and financial backers, may find it useful to define a second type of service life. This is the functional life, the period for which a building serves its intended purpose.[3] DfD is concerned with extending a building's functional life through adaptability and also, by extension to the life of its components, through deconstructability and reusability.

In practice, the technical life of a building usually exceeds its functional life. Designers may forgo DfD when the opposite is true: monuments, civic structures such as museums, and transportation structures like bridges and tunnels nearly always require replacement or rehabilitation before their usefulness expires. But the vast majority of buildings—commercial properties in particular—become ineffective and unwanted while still structurally sound. Such buildings are usually demolished to make way for new construction; a 1990 study set the average lifespan of a building in the United States at less than thirty-five years.[4] Certainly, there are "icons" that physically outlive their functional lives. But a pretty building in which nobody can live or work misses the point of architecture, and most modern architects would prefer their creations to survive as productive spaces, not as artwork. Buildings are meant to be occupied.

One may find a clear example of this issue in the works of architect Frank Lloyd Wright. Like many other architects of his time, Wright designed for perfection, specifying virtually every detail of his spaces, down to custom windows and furniture. Wright's fame and ego produced inflexible buildings that could never adapt to societal changes. Perhaps the best instance is the Johnson Wax Building, a rigid enclosure loosely based on the structure of a tree and completed in 1944 in Racine, Wisconsin. Less than a decade after its completion, new fire-safety codes condemned the building due to its single central stairwell; with no possibility of horizontal expansion, the tower has stood unused ever since.[5] Likewise, Wright residences were always untouchable and consequently impossible for most people to live in comfortably, strictly limiting their functional life. Though the works endure as historic landmarks, they no longer serve any purpose except as massive sculptures, museums dedicated to themselves.[6] Worst of all, when one building becomes functionally dead, somebody must build another, to accommodate the displaced tenants. Icons, then, are no better than conventional buildings that get torn down: not sustainable at all.

Johnson Wax Building

A particular mindset is to blame for the proliferation of useless structures, a trend promulgated largely by the "starchitects" and master buildings of the early-to-mid twentieth century. These architects were likely to treat their end product, the complete building, as a fixed entity—something that would never change once complete. As a result, construction practices developed such that today's conventional buildings are characterized by complex, integrated structures with a high interdependence of all systems, structural and otherwise. This approach makes reuse impossible, leaving recycling as the "greenest" possible recourse and demolition the most likely end-of-life scenario. DfD seeks new design and construction techniques that improve the capacity for end-of-life salvaging through front-end detailing.

How do architects come to terms with the fact that their works will one day be torn down? Surely, for every Wright, there are a thousand architects for whom not a single work will survive the next century. But these architects can make a similar bid for immortality by designing maximally adaptable structures, so the buildings live out their full technical lives serving any number of different purposes. They can do one better by incorporating the principles of deconstructability and reusability, so the building physically lives on in future structures—a potentially infinite cycle. It is to the architects' advantage to learn how to apply these principles without sacrificing form.

Building-Level Strategies

Webster and Costello have compiled a list of general strategies that can be applied toward building design to improve the potential for successful disassembly.[7] This list may be stratified into the building level, the system level, and the component level.[8] The building level involves design decisions that consider the building as a whole, to increase the likelihood that future developers will attempt adaptation or deconstruction. This choice will likely be financially motivated, so disassembly must be economically competitive with straight demolition. One essential strategy toward proving the worth of disassembly is to encourage engineers to develop a physical deconstruction plan as part of the construction documents.[9] Such a plan could include a list of building elements with their technical lives and potential for reuse, as well as instructions, such as sequencing information, to aid in taking the building apart. Related strategies include safeguarding as-built drawings in a secure on-site location and labeling components whenever possible.[10] Encouraging this sort of documentation might propel engineers and architects to think more seriously about deconstruction and develop their own creative solutions.

Perhaps even more important than preparing the right documentation is selling the concept of DfD to owners and financial backers. Certainly, DfD will be considered only by an owner who believes its principles to be profitable in the long run. Designing for adaptability improves the property value (and resale value) because the existing spaces may be reconfigured for the needs of future tenants and owners. Designing for deconstructability reduces future liability, as demolition of the building could be not only expensive but difficult on the community.[11] And finally, the rising cost of energy may well convince owners that designing for reusability is the best way forward. It is energy efficiency that has made steel recycling the rule rather than the exception: steel can be melted and re-formed using only about 25 percent of the energy input needed to mine iron ore and produce new steel.[12] But DfD goes one better: if an element is simply salvaged, and reused "as is," essentially the only energy input required is the cost of transportation to a new site.[13]

System-Level Strategies

System-level strategies define the various systems of a building (structural and otherwise) to make them as easy as possible to reconfigure or take apart. Two of the most effective system-level strategies are transparency and regularity. Transparency of the structural system means making structural members visible and connections accessible, helping a dismantler determine how to proceed with disassembly.[14] It also means keeping structural and mechanical systems separate, so that utilities threaded through girders and the like do not impede renovation or deconstruction. Regularity means the use of repeating structural systems, with a relatively small palette of different parts. This technique improves adaptability since regular systems may be added, removed, or reconfigured without modifying the entire structure. Furthermore, it allows dismantlers to expect certain conditions and enables them to sort members and connections after taking the building apart. More complex designs that specify structural parts as an afterthought to architecture typically result in a great variety of sizes, angles, and material grades, which makes a dismantler's job difficult.[15] Architects and engineers alike, with an increasing regard for sustainable designs, must trend toward simplicity.

The mills and factories built throughout the northeastern United States during the nineteenth-century industrial revolution embrace these tenets perfectly. These brick-and-timber or brick-and-concrete structures proliferated and then lingered when their industries moved elsewhere. Some were torn down; in these instances, the regular layouts and soft lime mortars between bricks made disassembly and material salvage more economical than demolition. Others were restored as offices, warehouses, restaurants, and homes, such as the East Boston, Massachusetts, factory shown. The regular layout of some buildings is eminently adaptable, allowing conversion to condominiums without changing the basic structural system. To be fair, one further characteristic of these nineteenth-century structures enables their successful reuse: charm. Though they were erected

Converted East Boston Factory. Exterior

Converted East Boston Factory. Typical Condominium Unit

as cheaply as possible to serve their intended purpose and with little regard to aesthetics, the buildings today evoke the quaint lifestyle of a bygone era. In modern developed countries, the building techniques used to build these factories are unusual: bricks, for instance, are applied as afaçade but rarely as a load-bearing wall.[16] Therefore, the exposed bricks and concrete mushroom columns lend real estate value to the East Boston condominium units shown, and these components have a high salvage value as well. It is hard to imagine that a person could ever feel a similar nostalgia for a hunk of steel, but one must distinguish between "sentimental" salvage value and structural salvage value. In the correct market, any element that can be detached from a building still in good condition has a potential for reuse.

Component-Level Strategies

Finally, component-level strategies involve making individual elements of the building easy to separate. Proper connection design in particular can make the difference between a component that gets scrapped and one that is salvaged for reuse. Economically, connections represent up to half the cost of construction and are responsible for most design, fabrication, and erection issues.[17] Moreover, connection design may dictate the condition of the structural elements—columns and beams—when they are salvaged for reuse. Most connections use either mechanical fixings (such as bolts) or chemical fixings (such as welds and construction joints) or a combination. A chemical connection is difficult to remove cleanly. A me-

chanical connection is more desirable, though it often leaves holes that require patching to make the member complete again. Some connections require elements to be coped, or trimmed to fit in a certain position, while others keep elements intact. Accessibility of connections is another major issue that determines how easily a structure may be dismantled piece by piece. The challenge to engineers and architects is to satisfy strength constraints while minimizing damage to the elements.

Numerous component-level strategies are showcased in a typical IKEA retail facility.[18] The basic structural system is a steel column and open-web truss system. Interiors are left unfinished, with structural members and connections exposed for easy identification and access. (The interiors simultaneously embrace the system-level strategy of defining spaces with nonstructural, reconfigurable partition walls.) In addition, IKEA facilities are built with a slab-on-grade spread footing where possible, which is not only cheaper to construct than a deep foundation but also permits faster, more economical deconstruction. Drilled shafts and driven piles, which are difficult to remove and almost impossible to reuse, are avoided in favor of this shallow foundation.

Returning to the above ground portion of the structure, it may be noted that steel has an exceptionally high potential for reuse among the

Exposed Structural System in an IKEA Retail Facility

standard building materials today. Individual steel members are versatile and durable. Extruded members are factory-made with low tolerances to a standard system of shapes, making it relatively easy to match a salvaged member with a buyer. Today, 11 percent of the world's structural steel is reused after demolition. Of the remaining 89 percent, 87 percent is recycled and only 2 percent is sent to landfill.[19]

One limitation of steel is its tendency to rust and deteriorate over time. If a bolted connection becomes rusted into place, disassembly will be hindered and may cause damage to the members, diminishing their usefulness as reusable parts. The same is true for clamped friction connections. Perhaps for this reason, a specialty steel system developed in the early 2000s, Quicon, has not received widespread market attention despite its proven ease of assembly. Developed by the Steel Construction Institute, a consultant based in the United Kingdom, Quicon furnishes steel beams with integral shoulder bolts, and T-brackets to create connections. The T-brackets include a series of keyhole-shaped notches that slide onto the shoulder bolts on-site.[20] Unfortunately, the main advantage of this system—a connection that can be deconstructed without damage to the members—may be overshadowed by an expectation that the connection will rust in place and defeat its purpose. With any system, designers can ease the threat of steel deterioration by specifying galvanized steel, cladding the members, or otherwise protecting structural components from the elements. This principle applies to all steel construction, of course, but it is particularly relevant to DfD.

Concrete has a far lower potential for reuse than steel. There is no way to reorganize or reuse pieces of a cast-in-place concrete structure, especially a reinforced concrete structure with a continuous rebar cage (virtually all concrete structures are reinforced in this way), because there are simply no distinct pieces. Even recycling concrete is not economically viable, at least in the usual sense of recycling. After a concrete building is demolished, or waste concrete from a new construction is broken down to be hauled away, whatever tensile strength the concrete had between particles is effectively gone. Concrete made from recycled aggregate is not as workable or durable as concrete made from virgin aggregate, yet it requires the same input of Portland cement, so decreases in energy input required or carbon dioxide released are trivial.[21] The best possible end-of-life scenario for cast-in-place concrete is probably "downcycling" to a fill or road base, which is better than a landfill but certainly does little to close the materials loop.

There is hope, however, for precast concrete. In contrast to cast-in-place concrete, precast concrete may be produced in regular-sized individual pieces that can potentially be detached from one another during disassembly, like steel members. One way to ensure precast systems are deconstructable is by connecting the pieces with mechanical fasteners such as anchor bolts, leaving room for thermal expansion to allow cracking.[22] Another tactic is to use pieces that naturally interlock, like tunnel forms, thereby avoiding the need for any additional fixing at all. The relatively common practice of casting corbels into concrete columns, for example, allows almost a gravity support system for floor joists and roof purlins. No matter what material is used for the structural system, it is evident that architects must work closely with engineers to implement the principles that make DfD a reality.

Conclusion

The principles of DfD suggest many strategies for enabling the disassembly of a building through decisions made in the design phase. These decisions depend on the architect maintaining a dialogue with the structural engineer and the service engineers. All parties will be required to make concessions. The structural engineer might need to settle for a load-carrying system that is far from optimal—specifying a heavy bolted connection where a simple fillet weld would otherwise do the job; specifying an extended connection where a regular one would require coping of the beam flanges; perhaps specifying larger members than necessary to reduce the need for stiffeners. The service engineers must agree not to follow the standard practice of threading utilities through structural components, and instead to keep all systems independent to facilitate disassembly. Most of all, the architect must deal with a new set of constraints on what forms are physically possible, fulfilling each project's purpose yet allowing for adaptation or deconstruction should that purpose change.

But restrictions breed creativity, as the saying goes, and in close collaboration the architect and the engineers may discover elegant solutions that would never have been dreamed of otherwise. The gap between vision and reality, between beautiful, functional spaces and transparent structural systems easy to take apart, can be narrowed. The additional costs associated with separating building systems and with the actual process of dismantling can be reduced, to the point where DfD is competitive or even more economical than conventional design. Only time will tell how DfD will be implemented in a world of rapidly dwindling resources, but it is the duty of the architect to take the initiative to reduce the building industry's vast energy requirements through emphasizing the simple principle of reuse.

Notes

1. Asko Sarja. *Integrated Life Cycle Design of Structures* (London: Spon Press, 2002), p. 19.
2. Mark D Webster and Daniel T Costello. "Designing Structural Systems for Deconstruction: How to Extend a New Building's Useful Life and Prevent It from Going to Waste When the End Finally Comes" (Proceedings of the 2005 Greenbuild Conference, Atlanta, November 2005), p. 1.
3. Elma Durmisevic and Jan Brouwer. "Design Aspects of Decomposable Building Structures" (Proceedings of the Deconstruction and Materials Reuse Conference, Karlsruhe, Germany, 2002).
4. Sarja. *Integrated Life Cycle*, p. 21.
5. William Allin Storrer. *The Architecture of Frank Lloyd Wright: A Complete Catalog*, updated third edition (Chicago: University of Chicago Press, 2007), p. 6.
6. Stewart Brand. *How Buildings Learn: What Happens after They're Built* (New York: Penguin Books, 1994), p. 58.
7. Webster and Costello. "Designing Structural Systems for Deconstruction." Durmisevic and Brouwer. "Design Aspects of Decomposable Building Structures," p. 20.
8. Webster and Costello. "Designing Structural Systems for Deconstruction," p. 11.
9. Ibid., p. 6.
10. Bradley Guy and Nicholas Ciarimboli. "Design for Disassembly in the Built Environment," City of Seattle and Pennsylvania State University, p.9.

http://your.kingcounty.gov/solidwaste/greenbuilding/documents/Design_for_Disassembly-guide.pdf (accessed 17 June 2011).
11. Steel Recycling Institute.
12. Mark D. Webster. "Structural Design for Adaptability and Deconstruction: A Strategy for Closing the Materials Loop and Increasing Building Value" (Proceedings of the 2007 ASCE Structures Conference, Long Beach, CA, 2007), p. 1.
13. Webster and Costello. "Designing Structural Systems for Deconstruction," p. 3.
14. Ibid., p. 4.
15. Ibid., p. 6.
16. Graham W Owens and Brian D. Cheal. *Structural Steelwork Connections* (London: Butterworth, 1989), p. 128.
17. Guy and Ciarimboli. "Design for Disassembly," p. 34.
18. Elma Durmisevic and Nico Noort. "Re-use Potential of Steel in Building
19. Construction" (Proceedings of the CIB TG 39–Design for Deconstruction and Material Reuse Conference, Delft, Netherlands, 2003), p. 3.
20. "Steel Construction Institute," http://www.steel-sci.org (accessed 8 May 2009).
21. Lauren Kuntz. "The Greening of the Concrete Industry: Factors Contributing to Sustainable Concrete," master of engineering thesis, MIT, 2006), p. 3.
22. Webster and Costello. "Designing Structural Systems for Deconstruction," p. 10.

Design with Climate: The Role of Digital Tools in Computational Analysis of Site-Specific Architecture

Azadeh Omidfar and Dan Weissman

When your house contains such a complex of piping, flues, ducts, wires, lights, inlets, outlets, ovens, sinks, refuse disposers, hi-fi re-verberators, antennae, conduits, freezers, heaters–when it contains so many services that the hardware could stand up by itself without any assistance from the house, why have a house to hold it up? When the cost of all this tackle is half of the total outlay (or more, as it often is) what is the house doing except concealing your mechanical pudenda from the stares of folks on the sidewalk?

–Reyner Banham, "A Home is Not a House" (1965)[1]

The discourse on semantics of "vernacular" is inconsequential. What matters is the localized knowledge base of climatic responsive formal techniques/operations at scales ranging from the body to the city that have been underutilized or "lost" since the advent and subsequent reliance on "active" systems, or those that rely on electrical energy to produce environmental effect.

–Amos Rapoport, "Defining Vernacular Design" (1990)[2]

The proliferation of new digital tools for computational analysis in parallel with improved fabrication techniques enables the potential for radically reinvented passive architectural systems, closing the long-standing schism between architectural and engineering design processes. These techniques may be based on either vernacular methodologies or wholly new forms yet untested. For example, the Hewlett Packard Software Campus in Bangalore, India, designed by Rahul Mehrotra Architects, and the DSW Building, by Sauerbruch Hutton in Berlin, both critically employ digital climate-design tools to create novel formal architecture. They employed "high technology" in the design phase to engage architectural form as the climate systems, and thereby reduced the necessity of mechanical/electrical technologies during the lifetime of the buildings. In these cases architecture can be seen, from the moment of inception, as a technological device able to employ basic principles of physics in the design and construction of a finished architectural product. Through well-integrated design processes, such elemental principles as aerodynamics, optics, and thermodynamics can be integrated to "thicken" formal strategies and spatial organizations, instead of layered on as discrete systems, as lamented by Banham.

One measure of contemporary architectural design is its effectiveness in creating positive environmental changes, whether the "silver bullet'" of CO_2 emissions reduction or decreased dependence on fossil fuels and associated lowering of energy costs. Although energy-saving techniques contribute to both social and economic sustainability, here we shall sideline this critical discussion. It is our responsibility as designers to understand and engage sustainability metrics through design thinking, questioning reductive assumptions about the world while promoting radically new methods of evaluation. However, it must first be understood that this contemporary drive is not new but has significant roots in recent history.

A False Modernity

Originally employed to shelter humans from the natural world, building was devised as a means of protection from uncomfortable weather conditions and the intensity of natural disasters. Climatic responsiveness was deeply embedded in architecture through locally understood collective knowledge. Western industrialization, and with it the development of active systems that deliver heating, cooling, and artificial light, has decoupled architecture from context. Glass towers in the desert proliferate.

To emphasize this distinction, many Modernist architects sought to create a universal language for architecture that could transcend particular site conditions by celebrating the mechanistic and the technological. A reliance on active technologies such as HVAC systems and fluorescent lighting allowed the new architecture to transcend space and time, and rather than relying on local context, to become an "International Style."

This line of architectural Modernism explicitly opposed vernacular or regional approaches to architectural production. However, a less-discussed strand of Modernist thinking serves as the basis for today's digital tools: the pioneering research into climatically responsive architecture at many leading Western architectural institutions throughout the 1950s. Work by Victor and Alday Olgyay at Princeton, MIT's Solar House Project, and the Architectural Association's research into hot and humid climates provided the basis for many computational analysis programs.[3] Victor Olgyay's treatise provided a set of tools with which build-

ings could be more directly tailored to their local climates. Ecotect, an environmental simulation software, was developed from the Olgyay brothers' work, providing a digital graphical user interface and algorithmic structure to a set of concepts originally published a half-century earlier.

Since the 1950s, funding for research on climatically responsive architecture, and the resulting academic coursework required of students, has ebbed and flowed with global energy politics. Yet the frequency and intensity of recent extreme climate events, the growing human population, and continued dependence on diminishing fossil fuels have heightened a sense of urgency toward sustainable design and construction.

Models

Ignasi de Sola-Morales, in his essay "Weak Architecture," discusses an archeological "deconstruction" of layering that exposes more information about the site and uncovers an additional tool for describing a superimposed reading of reality.[4] To design a structure at a particular location, a kind of "deconstruction" must occur to gain a deeper understanding of the "layering" of the site and its forces. Today's digital tools provide designers access to the collective knowledge accrued by humanity of the immaterial forces at work in our world, allowing a richer and deeper understanding of specific sites and contexts. With this new power, however, it is imperative that users in both industry and academia have a thorough understanding of three principles:

1. Computational models are often simplifications of reality for pragmatic purposes, and simplifications inherent in those models may be overlooked by users.
2. In an attempt to ease their use, certain programs provide default settings that may not be appropriate for every project in every environment.
3. Available data, such as for weather or electrical lighting files, may be not fully accurate for the particularities of any given project.
In parallel with other facets of the architectural discipline, such as structural design or building

fabrication, the skill set for analyzing and designing low-energy-consuming buildings was long ago forfeited by architects and handed off to engineers. However, a wide range of digital tools such as Ecotect, Transys, DesignBuilder, ESBR, and IES have begun to reverse that trend by providing architects (and engineers) with highly articulated models of how designs may perform. These tools allow for a generation of output diagrams for many elements of climate-responsive design including solar radiation, daylighting, natural ventilation, and user comfort.

For better or worse, these output diagrams, graphs, and images are often assumed to be scientifically accurate despite two underlying issues. First, digital tools rely on mathematical models, which must externalize certain elements of reality, such as complex thermodynamic phenomena, either for computational purposes or so that a cause-and-effect relationship may be identified. In physical reality, however, these externalities may be defining drivers of environmental conditions, and simplification may cause design decisions to be based on completely incorrect assumptions. Second, a model's accuracy is only as good as the base information provided from the start. Incorrect assumptions or default settings may dramatically alter results. Therefore it is critical for the designer to understand the limits of each software package before using it as a design tool. We shall investigate here the shortcomings of two widely used simulation tools: Ecotect and DesignBuilder.[5]

Simplified Algorithms

Although Ecotect allows for many types of analysis, its most widely used function is probably daylight simulation. Offering a quick interface for understanding sun position and shadows, Ecotect also performs basic calculations such as daylight factor (the percent of daylight inside a structure as a function of building geometry). However, Ecotect's internal daylight calculation engine uses the "split-flux method" algorithm, a highly simplified calculation that calculates only direct light and a first bounce. It does not, however, calculate any inter-reflected light from surfaces in the environment or building space, often called global illumination or ambient bounce in other software packages. Because of this simplification, the simulation has the significant potential to provide highly incorrect light-level indications. Design decisions based on such results may fundamentally alter the building's performance, which could include heavy contrast ratios and glare or unintended dark spaces where electric lighting must be employed to make up the difference. Ecotect's developers were acutely aware of this limitation and included the option to export building geometry and materials into a more robust software tool, the validated Radiance engine, for accurate lighting analysis.

The inadequacy of Ecotect's internal daylight simulation manifested itself in the recently completed Heelis National Trust Central Office in Swindon, United Kingdom, designed by Feilden Clegg Bradley Studios. When lighting analysis was performed in Ecotect, problematic areas were shown brighter in the simulation than in the actual built design, resulting in the unintended overdependence on electric lighting. Although the building did not achieve its ambitious performance goals, it still managed to outperform most buildings in the United Kingdom.[6]

The problem of simplified calculation algorithms also arises in DesignBuilder when generating simulations for thermal mass. Unlike lighting calculations, thermal calculation algorithms are not directly based on physical geometry. In the calculation, DesignBuilder distributes solar radiation into digital space based on simplified window-to-wall ratios and U-values for material assemblies, disregarding building geometry. Although this method works well for calculating internal loads in buildings, designing effective thermal mass requires an additional direct solar radiation calculation, which must be based on exact building geometry and site location. When performing a thermal mass calculation in DesignBuilder, users are presented with options for calculation complexity. The simplest and fastest calculation combines all windows into a single window-to-wall ratio and disregards direct solar. This calculation can provide only ballpark figures, as it will always underpredict energy savings. The more complex option, exporting precise window locations for an added direct-solar calculation, takes longer but results in a more accurate calculation. Even this longer method produces inaccuracies, as DesignBuilder also uses the split-flux method to calculate daylight. To attain truly accurate results, the solar calculation would require the use of Radiance.

To complicate matters further, DesignBuilder offers a check-box for easily applying thermal mass. Instead of including this mass in architectural surfaces (which may be added manually when designating materials), this setting merely creates a block of concrete in the middle of each room that may absorb only conduction, not solar radiation. Because this block of thermal mass responds only to conduction, imprecise air infiltration rates could change the impact of air temperature. If one uses the default setting that assigns a single infiltration rate for an entire building, the same quantity of air is delivered from outside to every room of the building. This is an impossibility when interior zones exist that are not adjacent to exterior surfaces, producing vast inaccuracies across the calculation.[7]

Weather Files

Although not the only program able to load and manipulate weather files, Ecotect is widely used for weather analysis in the early stages of architectural design. To begin, one must download a local weather file from the U.S. Department of Energy (DOE) website. These "typical meteorological year" (TMY) weather files represent expected, but not actual, weather patterns at a given location, which include temperature, humidity, wind direction and speed, solar radiation, and rainfall. These files may be problematic for two reasons: location accuracy and file data accuracy.

First, DOE catalogues only climatic information from available weather stations, which until recently were mostly at airports, as flight required highly accurate weather prediction. Airports, however, are situated on large, open areas of land with climatic conditions far different from urban conditions. This can highly skew building-performance criteria when the geographically closest weather file to a project site has different micro-climatic conditions. Although solar radiation, rainfall, and humidity may not be radically different depending on local conditions, wind patterns are notoriously location-specific.

Second, TMY file data may not be an accurate representation of local climates. TMY files are generated from a thirty-year timeframe, where TMY = 1952–1975, TMY2 = 1961–1990, and TMY3 = 1976–2005. As better surveying techniques and climate change proliferate, older climate files cease to provide accurate pictures of their locations. Moreover, as the files are compiled from statistically defined weather patterns over the past century, they do not have the capability to determine future weather patterns as climate shifts occur. Although algorithms that can "morph" existing TMY files for projected climate change have been developed, such as the tool by the University of Southampton, the algorithm is based on 2001 climate change data, which is also outdated. The weather files attained through the DOE are the best we have, and although they will continue to be used for analysis, their limitations must be considered.[8]

Default Settings

Beyond data accuracy, incorrect use of weather file information may result in serious building performance losses. For example, when one opens the wind-rose dialogue box in Ecotect, the first default diagram shows wind over the entire year. Although orientation toward prevailing winds for natural ventilation is ideal, orientation must be considered seasonally depending on local climate, such as toward the prevailing wind in the summer, but not the winter, in northern climates. Programs such as Ecotect could assist future users in avoiding such basic pitfalls by modifying the way users access such presets. Ecotect also includes a building orientation optimization tool, which employs a black-box calculation based on solar radiation and, presumably, wind. Although the calculation seems

GUND HALL ELECTRICITY: Measured vs Simulation

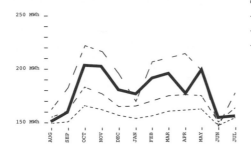

— Gund Measured 2007/2008

- - Simulation with Custom Inputs

- — Simulation with Default Occupancy & Plug Loads

···· Simulation with Default inputs

Measured vs. simulated electricity consumption,
Gund Hall, Harvard Graduate School of Design.

GUND HALL HEATING

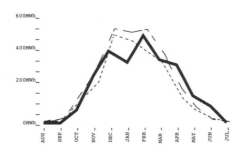

— Gund Measured 2007/2008

- — Simulation with Custom Inputs

- — Simulation with Default inputs

Measured vs. simulated heating loads, Gund Hall,
Harvard Graduate School of Design.

GUND HALL COOLING

— Gund Measured 2007/2008

- — Simulation with Custom Inputs

- — Simulation with Default inputs

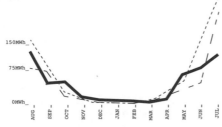

Measured vs. simulated cooling loads, Gund Hall,
Harvard Graduate School of Design.

All Custom Inputs	Electricity	Heating	Cooling
Annual % Difference	0.2%	8%	-5%
Mean Bias Error	0.002	0.04	-0.13
Root Mean Square Error	0.233	1.67	3.83

Simulated vs. Measured Data			
All Custom Inputs	Electricity	Heating	Cooling
Annual % Difference	-18%	-9%	43%
Mean Bias Error	-0.174	-0.13	-25.54
Root Mean Square Error	0.639	1.46	5.18

Simulated vs. measured data, summary tables.

to be based on the Olgyays' calculations for optimizing orientation, Ecotect does not provide the opportunity for designers to understand the basis for this calculation. If based on specific weather file data, this tool too may have its limits.

A more extreme example of this problem is revealed in DesignBuilder. In an attempt to create a user-friendly environment for beginners, the program initiates a set of default values when a building form is created. These default values are independent of the project, site, and climatic conditions. As many of the presets are hidden in dropdown lists, default parameters (such as the thermal mass calculation) may be easily overlooked, resulting in inaccurate design decisions. Holly Wasilowski Samuelson and Christoph Reinhart investigated the difference in simulation results using both default values and custom values in "Modeling an Existing Building Using Customized Weather Data and Internal Load Schedules as Opposed to Default Assumptions—A Case Study." [9] The authors analyzed Harvard Graduate School of Design's Gund Hall to understand the difference in results. The graph shown illustrates the discrepancies.

The tables shown illustrate the error in simulation results when compared to measured electricity,

heating, and cooling utility data provided by Gund Hall's building manager, shown by the thick blue line. The top dashed line shows simulation results using all custom inputs. The other two dotted lines show simulations using various combinations of default inputs.

The table on the top shows the "All Custom Inputs" results and the bottom table shows the "All Default Inputs" results. For all electricity, heating, and cooling, custom values clearly represent reality closer than default settings. For example, the annual electricity consumption is off by 0.2 percent for the custom simulation and 18 percent for the default simulation.

It is understandable that programmers include presets in such analysis tools, as many architect users are not well versed in rule-of-thumb settings. If simulation tools required all parameters to be input by the user, many designers simply would not spend the time creating the simulation. But by providing presets, simulation tools unintentionally introduce a stumbling block to accurate simulations. A balance must be struck that allows designers to gain an intuitive understanding of all parameters and appropriate values for their design and context, without sacrificing the accuracy of calculation.

Conclusion

Buildings are significant consumers of resources, and we desperately need for designers to begin weaning buildings off fossil-based energy sources. But in this shift toward sustainability, climatically responsive architecture has failed to reengage architectural discourse since its moment in the history of Modernism. Sustainability today is too often merely an appliqué of technology intended to sell real estate or perform didactic roles, and not truly able to produce new architecture. Moreover, architects often lament the systematization of the building industry, and sustainability as the apex of prescriptive requirements. Digital tools offer the opportunity for a paradigm shift, allowing architects the potential to design novel form, materiality, atmosphere, and experience that critically engages any particular project's full context while furthering our discipline.

Designers should note, however, that venturing into this new world requires the telling of cautionary tales. Analysis tools are only as good as the ability of their users to apply them appropriately in the design process, and to input and interpret the data correctly. The aforementioned issues are not spe-

cific to Ecotect or DesignBuilder, as many other tools contain their own unique limits. Yet these tools are extremely powerful if used correctly and their limitations understood.

In the near future we will see continued advances in design integration, as well as computational algorithm accuracy and user interface improvements. Predictive climate/weather data for the entire earth will become available in real time, along with seamless interfaces that allow for continuous feedback throughout the design and fabrication process and through the lifetime of buildings. While promoting "design with climate," the academy must address potential pitfalls by training professionals through a holistic approach that integrates back-of-envelope calculations in parallel with digital tools, creating the atmosphere of healthy skepticism of all results. This suggests the need for radical educational shifts to foster such abilities within the designer: a masterful mediation between a trove of analogue and digital data and tools, with the critical understanding of when and how to deploy each in the design process. To design sustainably is to ask the right questions, understand site, and design with climate.

Notes

1. Reyner Banham. "A Home is Not a House," *Art in America* (April 1965), pp. 70-79.
2. Amos Rapoport. "Defining Vernacular Design," in *Vernacular Architecture: Paradigms of Environmental Response*, edited by Mete Turan (Aldershot, England: Avebury, 1990), pp. 67–101.
3. Victor Olgyay. *Design with Climate* (New York: Princeton Architectural Press, 1963); Barber, Daniel. "The Modern Solar House: Architecture, Energy, and the Emergence of Environmentalism, 1938–1963," PhD dissertation, Columbia University, 2010; Fry, Maxwell and Drew, Jane. *Tropical Architecture in the Dry and Humid Zones* (London: B. T. Batsford, 1964).
4. Ignasi de Solà-Morales Rubio. "Architettura debole = Weak architecture," *Ottagono* 92 (September 1989), pp. 87–129.
5. DesignBuilder is an energy analysis package developed as a front-end graphical interface for the Department of Energy's EnergyPlus. Ecotect, developed by Square One! and now owned by Autodesk, is a sustain-

able design platform for thermal, solar, and acoustic analysis and design, adopted by numerous architectural firms and taught in many academic institutions since its development in the mid-1990s.
6. Ian Taylor, course lecture for "In Search of Design through Engineers," Harvard University Graduate School of Design, 2009.
7. Diego Ibarra. "DesignBuilder," May 17, 2011. E-mail. 31 May 2011. See also Diego Ibarra and Christoph Reinhart. "Daylight Factor Simulations—How Close Do Simulation Beginners 'Really' Get?" Harvard University Graduate School of Design, 2009.
8—Seth Holmes. "Climate Change Risks from a Building Owner's Perspective: Assessing Future Climate and Energy Price Scenarios," master's thesis, Harvard University Graduate School of Design, 2011.
9. Holly Wasilowski and Christoph Reinhart. "Modeling an Existing Building Using Customized Weather Data and Internal Load Schedules as Opposed to Default Assumptions—A Case Study," Proceedings of Building Simulation 2009, Glasgow, July 2009.

Masdar Institute of Science and Technology

Foster+Partners / AKT

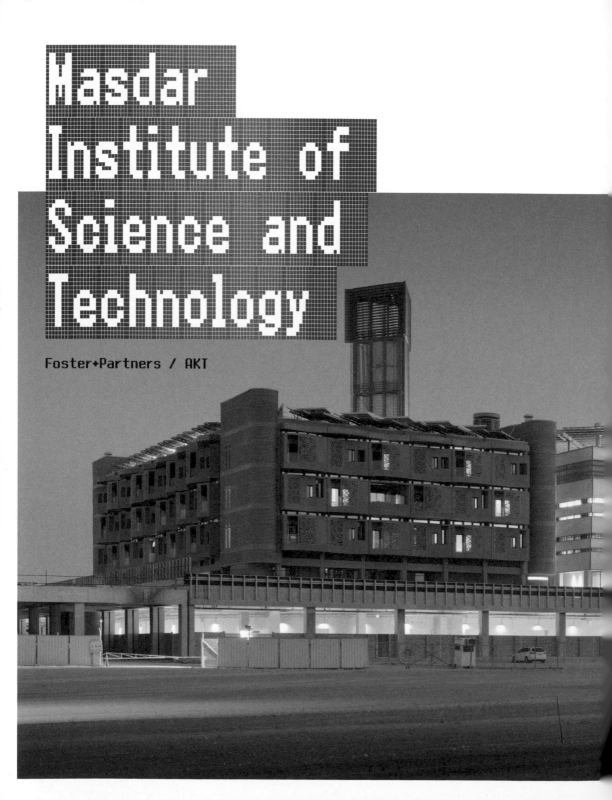

The Role of the Client in the Design of Sustainable Developments

Chris C. L. Wan

The start of the design process is usually signified by the execution of the design contract, followed by the design kick-off meeting. Everyone is introduced and the protocols for the design process are communicated, setting the ground rules. The client describes the project, sometimes provides a budget, and asks the design team to come back with a concept for approval. This process may suffice for traditional developments, but is not adequate for the successful design of sustainable developments. Given the growing awareness of the potential impact of climate change, the transformation from traditional developments to sustainable developments has begun and is an issue of increasing focus in the public domain. To facilitate this transformation, the client has a critical role to play to ensure that sustainable developments are actually sustainable.

To increase the chance of achieving sustainable designs, the design process for the client starts well before the kick-off meeting. During this phase the client needs to set the brief, determine the goals, calculate the budget, assemble the design team, and set up an integrated design process environment. The role of the client in this respect is pivotal, enabling the whole design team to operate as a collaborative entity to achieve successful delivery of the design services.

Clients come in many forms, from a single person with no building background to a property developer who employs a full suite of professionals. This essay is focused on the latter group of clients whose in-house professionals offer the potential to proactively contribute to the design process. Furthermore, I write from the perspective of building design to give my analysis a specific frame of reference.

Setting the Brief

Without an end user, there would be no end product. Therefore one must start with the end in sight and develop a brief that describes the requirements of the end user. It should be noted that the end user may or may not be the client, depending on the business model of the project. The physical requirement of uses, functions, and dimensions provides the basis from which the design team delivers a design fit for the given purpose.

Such is the advancement in the building industry that numerous requirements are governed by prevailing regulations and codes. These requirements include thermal comfort, ventilation, lighting, water supply, sanitation, use of safe materials, etc. Notably, these aspects reside in the environmental realm, which may give a clue to why environmental requirements are commonly not discussed or included in the brief. As a result, environmental requirements are often seen as minimum requirements that comply with codes and regulations. Sustainable development needs to look beyond the minimum environmental requirements and exceed them, either by expanding the scope of the brief to incorporate them or by determining a set of goals specific for the purpose.

Determining the Goals

As in many areas of life, Pareto's 80/20 principle is useful when setting goals with a sustainable agenda. The principle is to determine how to produce the most effective goals with limited physical and financial resources. Maximizing output from minimum input echoes ideas of efficiency and performance. Indeed, sustainable buildings and high-performance buildings share the same parameters: environmental, economic, and social factors.

Many design elements reduce energy demand with little or no capital cost, including building form, orientation to the sun, prevailing winds, shading and microclimate impacts. Planning efficiencies need to minimize unused spaces that otherwise would create an unnecessary drain on energy. Even the color palette should be considered, not only from the point of view of reflecting radiant heat but also from a maintenance aspect. If fewer cleanings are needed to keep a building looking clean, then less energy would be used.

Water demand reduction is most easily achieved through low-flow fixtures and fittings. More challenging is the quest to design out water leakages from the system, since leakage is a major contributor to per capita water consumption. The role of the client in defining the project brief and leading its effective implementation remains critical to tackling such challenges.

Construction waste is normally seen as an issue for contractors. However, the design team can take a proactive role in designing for minimum waste. Collaboration with the available supply chain for façade systems can inform the appropriate dimensional modules. A single design module that is fractionally larger than the standard module or exceeds the maximum transportation capacity will necessitate additional cutting and joints. Applied thought at the right time in the design process will have many benefits. For many structures the major contributors to their embodied carbon come from

building materials such as concrete, steel, aluminum, and block work. Concentrating on the supply chain of these materials can lead to significant reductions in embodied carbon, with minimal cost impact.

Calculating the Budget

Calculating the correct budget ensures that the development is commercially viable. Commercial viability is the key to moving the design from the digital space within the computer to the real world. The complexities of sustainable design require that the designers understand the concept of the financial model, just as the financers need to understand the key technical factors that drive the design. Tackling these aspects in isolation will compromise the whole, and in more extreme situations, lead to design failure.

The current challenging economic environment means that an accurate project financial model is a higher priority than ever. However, sustainable development is not a fashion or a passing trend but something that be viable in all economic environments. For sustainable designs to be adopted, the design team has to demonstrate understanding of the economic drivers. Designs with significant reductions in their environmental footprints must also be commercially viable. Development lease rates are commonly benchmarked against the existing market, even against buildings built without a green agenda. Commercially viable construction budgets for sustainable developments therefore need to be as close as possible, if not comparable, to the market "non-green" buildings.

Assembling the Team

Depth of knowledge in any particular field is often the understanding of how much one does not know. Whether the design process is described as integrated, collaborative, synergistic, or holistic, the common factor is that no one person, entity, or discipline holds all of the solutions required to answer the complexities and sometimes contradictions inherent in sustainable design. The way forward must stem from a team effort. Building the right team is not simply a matter of assembling the best individuals from each discipline, just as the best orchestra is not necessarily comprised of the best solo musicians. Members must also have the ability to work as a team. Successful experience by team members working collaboratively in the past does count.

Scope documents for design services is the place to start and should take into consideration what is needed to put the right team in place. In a world of specialization, the client must bring the appropriate range of design disciplines to the table. The idea is to provide the team with the necessary expertise to make informed design decisions. A typical list may include the following design disciplines that could comprise the consultant design team:

- Acoustics
- Architecture
- Architect of Record, Engineer of Record
- Building Information Technology and Information and Communications Technology
- Building Management Systems
- Building Performance Simulation
- Cost Estimating, Quantity Surveying
- Electrical Engineering
- Facility Management
- Façade Engineering
- Fire and Life Safety Engineering
- Geotechnical Engineering
- Interior Design
- Landscape Architecture
- Lighting Design
- Mechanical Engineering
- Plumbing Engineering
- Project Control and Management
- Renewable Energy Systems
- Security Systems
- Signage, Graphics, and Wayfinding Design
- Specifications
- Structural and Civil Engineering
- Transportation Engineering
- Vertical Transportation Engineering
- Water Management and Recycling
- Wind Engineering

The consultancy design team will be supported by the client, which may have an in-house technical team. The required range and depth of the client's in-house technical team will depend on the goals of the client. For a sustainable goal, the technical team may extend beyond traditional architecture and engineering to include specialists in energy, water, waste, and supply-chain management. Assembling such a diverse team will produce a synergy that results in a whole that is greater than the sum of its parts.

Setting Up an Integrated Design Process Environment

Traditionally the building design process is linear. This applies to the way that design stages are set up—such as concept, schematic, design development, and construction phases—as well as to building design practices. The process is typically led by the architect, which worked well when building design involved presenting formalistic solutions to the brief. With increasing demand for sustainable design, however, this linear design process promotes a piecemeal approach regarding environmental building features. Architects are finding out that these features are either driving up construction costs or sending the design process back to earlier design stages, raising design costs. Either way, sustainable development is not so sustainable due to cost increases.

Building information modeling can be our guide to the future of building design processes. The main idea is to promote collaboration among different disciplines by allowing everyone to work in a single software model. Derivatives of the modeling can be analyzed and tested for their environmental performance, enabling environmental concepts and features to be considered earlier in the design process. In effect, it allows for an iterative design with quick feedback loops, thereby supporting sustainable development without additional construction and design costs.

Bridging the linear process and an iterative process to create a practical integrated design method is the starting point. Small steps ensure that the whole industry moves together. Workshops bringing together all design stakeholders—both design and client team—are necessary for major interdisciplinary issues. Lessons learned by the client team in previous projects should inform design discussions, to avoid "reinventing the wheel." Instead, that time could be invested in exploring relationships among the environmental, commercial, and social aspects of the design.

Under a linear process, the design team is led by the architect, and a formalistic design approach prevails. The traditional concept is judged on its response to the brief and aesthetics. Most of the time, the supporting cost-consultancy scope does not even start until the concept design is complete, and a cost estimate is then produced. Furthermore, other project goals are not even considered, leading to compromised solutions. This also can lead to the unfortunate scenario in which, during the later design stages, the client is responsible for altering the concept when value engineering or scope-cutting kicks in, due to cost overruns. An iterative process allows other disciplines to take the lead from time to time. The structural engineer can be asked what he or she would propose if structural efficiency is the priority. The building services engineer can be asked what he or she would propose if energy saving is the priority. Similar questions can be posed to each of the other engineers and specialists. The cost consultant should propose systems and materials from

a cost point of view. Even the client team can put forward their ideals as members of the design team. It is probably true that the architect remains at the center of the process; the difference is that he or she is also the facilitator of integration.

At Masdar City, to give a specific example, the client team is actively confronting the challenge of creating a sustainable future. The client starts the process by setting the brief, determining the goals, calculating the budget, assembling the team, and setting up an integrated design process environment. The design team is immediately followed by the client, which serves as a facilitator of and participant in the integrated design process. None of the described client and design team activities taken in isolation are new or groundbreaking, just as the various individual technologies required to achieve a sustainable future already exist. What can be explored and expanded is in the way that these various design team and client activities can be brought together, leading to new process models. Integration is not just for design processes but equally important for client processes. As such, innovation happens at the level of interdisciplinary integration, and not at the level of isolated activities.

Conclusion
In the sustainable development business, the economic, environmental, and social aspects of a project are often perceived to be at odds, leading to conflicting requirements. Many discussions and studies have been carried out on the subject of the cost of environmental sustainability, or rather, the premium of environmental sustainability compared to business as usual. During the current economic slowdown, many believe that environmental sustainability is not affordable. Yet these same people, outside of their business environments, would concur that steps should be taken to mitigate the possible array of negative consequences of climate change.

The challenge is not to give in to compromised solutions that deliver neither commercial viability, reduced environmental impact, or improved quality of life. The goal must be to bring these three aspects of a project together to enhance and improve one another. It is the search for a win-win-win solution that is the ultimate goal of Masdar City and similar future developments: to make sustainable development the business-as-usual of the future.

Open House

Cara Liberatore, Leslie Mctague, Anthony Sullivan

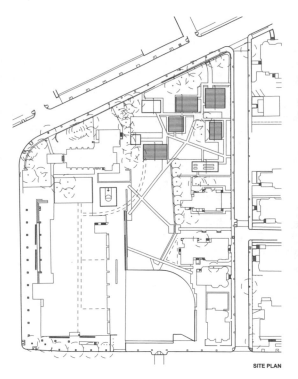

SITE PLAN

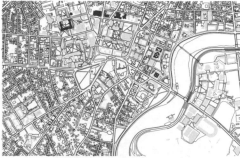

THE SCHEME INTENDS TO EXPAND ND BUILD ON THE CURRENT MUSEUM ALLEY IN THE OVERALL HARVARD CAMPUS.

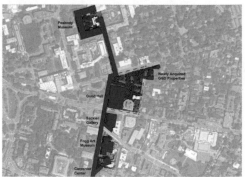

The Center for Contemporary Architecture is constituted as a series of pavilions arranged along a path between Gund Hall and recently acquired GSD properties across Sumner Road. The scheme seeks to create a new zone of activity within the growing GSD campus while creating an open and welcoming atmosphere for the public. The cluster of various-sized buildings is intended to be a place to share, discuss, and learn about architecture, an architectural open house. Each pavilion is surrounded by intimate outdoor spaces to allow activity to spill out from the galleries.

VIEW FROM GSD LAWN
BY BREAKING U P THE PROGRAM I NTO SEPARATE P AVILIONS THE
ACTIVE PATH BETWEEN THE OUTSIDE PUBLIC AND THE GSD LAWN IS
MAINTAINED AND ENLIVENED.

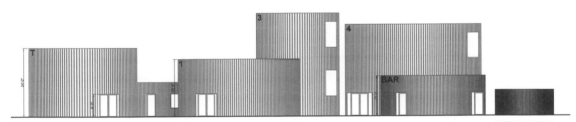

NORTH ELEVATION
SIGNAGE, GALLERIES ARE IDENTIFIED BY NUMBER, IS
BLENDED INTO THE CORRUGATION PATTERN IN A SIMLIAR
MANNER TO THE SIGNAGE AT 7 WTC.

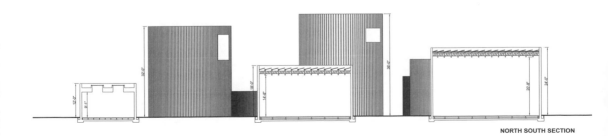

NORTH SOUTH SECTION

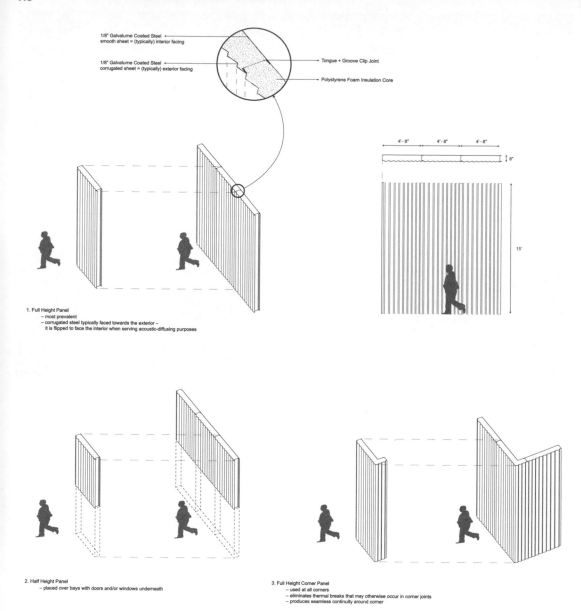

1/8" Galvalume Coated Steel smooth sheet = (typically) interior facing

1/8" Galvalume Coated Steel corrugated sheet = (typically) exterior facing

Tongue + Groove Clip Joint

Polystyrene Foam Insulation Core

4'-8"   4'-8"   4'-8"

8"

15'

1. Full Height Panel
   – most prevalent
   – corrugated steel typically faced towards the exterior –
     it is flipped to face the interior when serving acoustic-diffusing purposes

2. Half Height Panel
   – placed over bays with doors and/or windows underneath

3. Full Height Corner Panel
   – used at all corners
   – eliminates thermal breaks that may otherwise occur in corner joints
   – produces seamless continuity around corner

Each building is constructed of custom lightweight steel and structurally insulated panels. The structural panels allow for an extremely expedient construction process while providing a high-quality finished product. The insulated panels further provide high thermal efficiency, reducing heat loads in winter and cooling loads in summer. The buildings are each given a custom corrugation pattern, adding stiffness and stability while allowing for lyrical reflections of the surrounding environment.

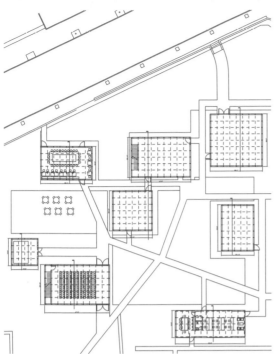

The gallery spaces are situated along the main path between the GSD and the intersection of Kirkland Street and Sumner Road with the bar and offices situated at opposite corners of the lot. The bar sits along the more public Kirkland Street with the offices situated along the quieter Sumner Road. The arrangement of the pavilions also creates a series of more intimate yards between them, to be utilized as an expansion of the gallery spaces on the interior. They are spaces to allow for discussion to spill out of the galleries in warmer months.

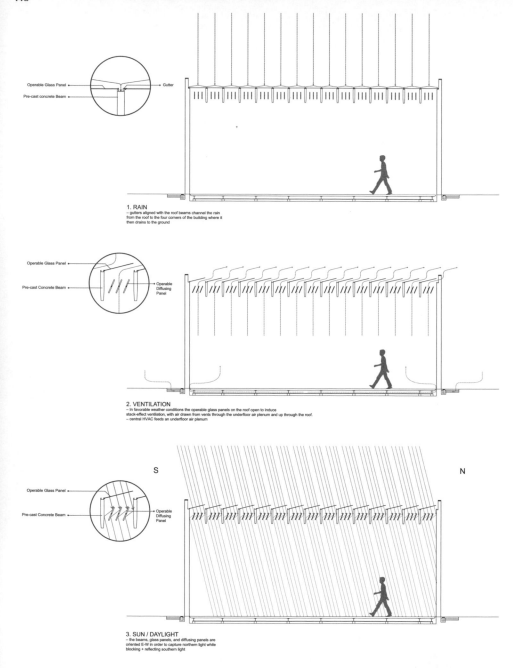

Operable Glass Panel —
Pre-cast concrete Beam —
— Gutter

1. RAIN
– gutters aligned with the roof beams channel the rain
from the roof to the four corners of the building where it
then drains to the ground

Operable Glass Panel —
Pre-cast Concrete Beam —
— Operable
Diffusing
Panel

2. VENTILATION
– In favorable weather conditions the operable glass panels on the roof open to induce
stack-effect ventilation, with air drawn from vents through the underfloor air plenum and up through the roof.
– central HVAC feeds an underfloor air plenum

S   N

Operable Glass Panel —
Pre-cast Concrete Beam —
— Operable
Diffusing
Panel

3. SUN / DAYLIGHT
– the beams, glass panels, and diffusing panels are
oriented E-W in order to capture northern light while
blocking + reflecting southern light

The galleries and offices utilize a system of adjustable louvers that allows for a controlled diffuse natural light to illuminate each gallery. Each pavilion also utilizes raised floor system coupled with a central HVAC system for all of the pavilions allows for an efficient delivery of heating and cooling in each pavilion.

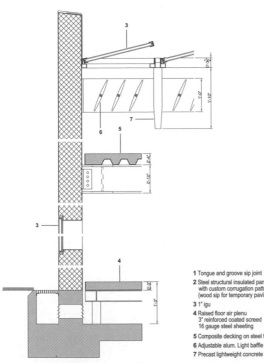

The Steel SIP construction allows for a tightly sealed building with minimal thermal bridging, providing a highly thermally efficient building, reducing overall heating and cooling loads throughout the year. Glazing is installed within each panel in the factory to provide a very high tolerance for glazing joints and sealing.

1 Tongue and groove sip joint
2 Steel structural insulated panel
with custom corrugation pattern
(wood sip for temporary pavilion)
3 1" igu
4 Raised floor air plenu
3" reinforced coated screed
16 gauge steel sheeting
5 Composite decking on steel framing
6 Adjustable alum. Light baffle
7 Precast lightweight concrete beam

SECTION DETAIL – EXTERIOR WALL

Refabrication

Charles Harris, Ji Seok Park

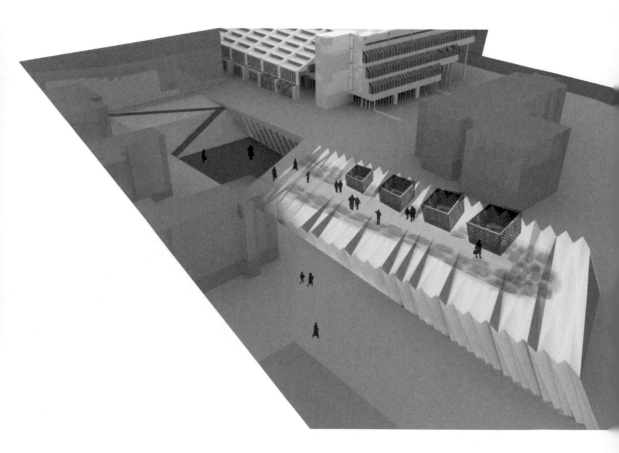

This project aims to reestablish fabrication as a primary component of architectural education. Currently fabrication facilities in multiple buildings at the Harvard Graduate School of Design occupy basement spaces cut off from classrooms, studios, and the public domain. This project proposes a sunken courtyard that links these basement facilities to a common space for the construction of large-scale models, prototypes, and temporary pavilions.

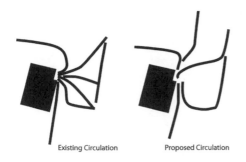

Existing Circulation   Proposed Circulation

At an urban scale, the building preserves circulation patterns around the courtyard while offering a new opportunity for dynamic outdoor space. The roof of the pavilion slopes with the interior space of the pavilion; as a result, it begins at ground level (adjacent to the courtyard) and rises to one level above ground (at street level). This sloped space could form an outdoor theater, sculpture garden, seating area, or event space facing the new central courtyard of the GSD campus.

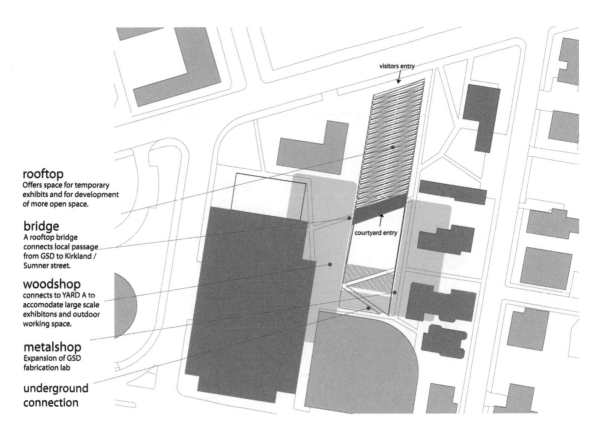

rooftop
Offers space for temporary exhibits and for development of more open space.

bridge
A rooftop bridge connects local passage from GSD to Kirkland / Sumner street.

woodshop
connects to YARD A to accomodate large scale exhibitons and outdoor working space.

metalshop
Expansion of GSD fabrication lab

underground connection

visitors entry

courtyard entry

This courtyard "plugs in" to existing basement facilities on three sides, connecting them with open space and natural light. On the fourth side, a sloped rectangular pavilion connects the sunken courtyard to the street, offering a new multilevel series of exhibition and gathering spaces that re-forms the GSD as a campus grounded in the making of the built environment and links its facilities with other Harvard schools and the local community.

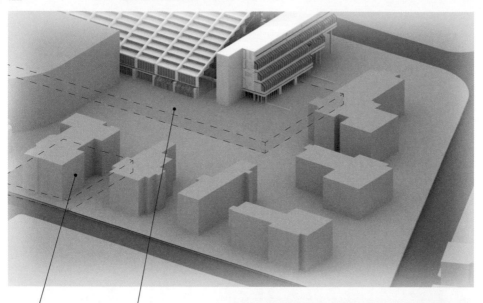

metal shop/cgis
Currently, these areas receive little to no natural light and are completely unknown to the general public.

fabrication lab
Currently, the GSD Fabrication Lab is tucked away in the basement, hidden from public view, bathed only in flourescent lights, and lacking large spaces for installations.

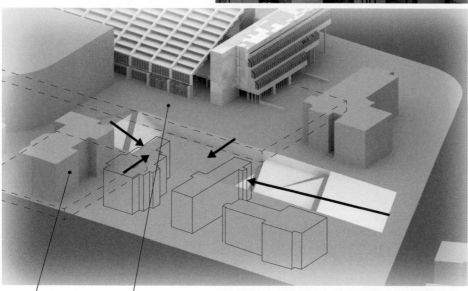

metal shop/cgis
From the CGIS library and the metal shop, the courtyard provides natural light and a connection to the community.

fabrication lab
The fabrication lab would have direct access to the courtyard, which would expose fabrication practices to the public eye and would supply large-scale spaces for installations and mockups.

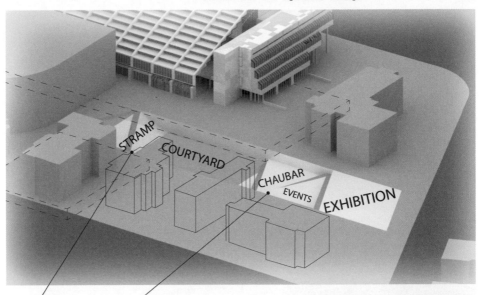

the stramp
To the South, the Stramp provides an integrated means of access to the courtyard. Additionally, the steps can serve as a venue for outdoor lectures, performances, or parties.

events/exhibition/chaubar
The Chaubar, events space, and exhibition space are envisioned as a series of connected platforms strung out between the courtyard and the street. Continuity of space allows flexibility for these programs.

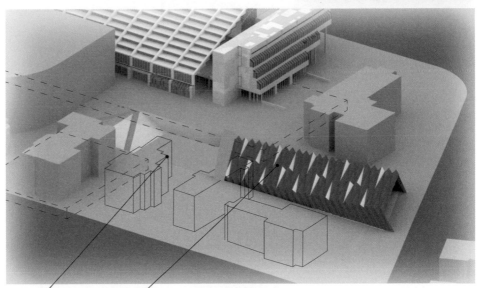

courtyard
From the courtyard, a glazed facade welcomes students into the Chaubar.

roofscape
The roof of the enclosure, accessible at ground level above the courtyard, offers the possibility for new types of outdoor social space.

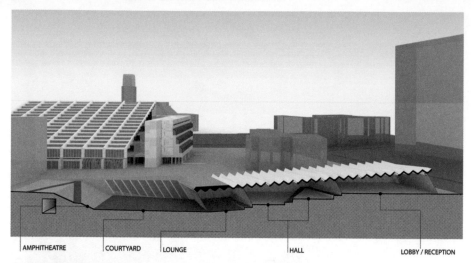

AMPHITHEATRE | COURTYARD | LOUNGE | HALL | LOBBY / RECEPTION

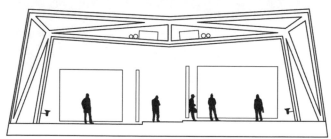

The sloped, rectangular, one-story pavilion provides the enclosed spaces envisioned in this proposal. This pavilion includes new spaces for the GSD community, including an expanded bar/gathering space (the "Chaubar"), an exhibition space, an office/meeting space, and a lobby. These spaces are divided by the building's main circulation corridor: a sloped walkway that zigzags across the rectangular volume connecting the spaces with the courtyard and the street level. As a result, the building programs occur at different levels as the ramp connecting them rises from the courtyard to the street.

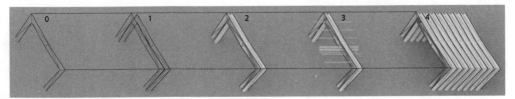

structural module
1 steel frame structure
2 insulation and interior and exterior panels
3 electricity / hvac / exhibition passes

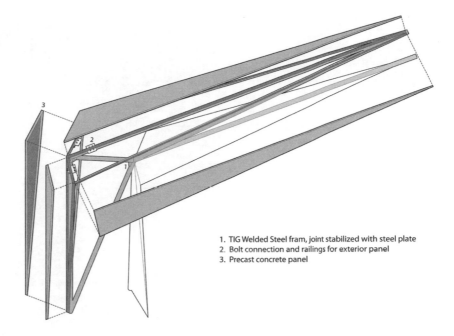

1. TIG Welded Steel fram, joint stabilized with steel plate
2. Bolt connection and railings for exterior panel
3. Precast concrete panel

Like Gund Hall, the new building would emphasize structural expression, to continue the theme of fabrication. However, rather than relying on a series of exposed trusses, this building adopts a modified folded plate system as a new means of structural expression. In this case, the structural steel members are not exposed; instead, they are covered with metal panels that follow the folded plate system while providing service areas required for the building's programs.

Ultimately, the aim of these new spaces is to reconstitute the GSD as a campus rather than just a building. And the space at the core of this campus will showcase fabrication facilities that emphasize the making of the built environment. In this way, the school can continue to be a renowned laboratory for architectural innovation, and it will expand its domain by providing a new platform for interdisciplinary research.

Heelis - National Trust Headquarters

FCB Studios / AKT

On Heelis: The Role Played by Environmental Engineers

Guy Nevill

Creating zero-carbon buildings and winning awards for innovation and sustainable design doesn't happen every day. More often, it comes when engineers, architects, and the design team work closely together, encouraging an open-minded and enlightened client. As environmental engineers, our role is to facilitate this—to push architects, clients, and designs as far as we can to achieving the best solution within the constraints of size, location, and budget.

Understanding of environmental and passive design varies across architectural practices. Some are well versed and great practitioners, their buildings influenced by a response to both the local and global environment. Others may be interested in sustainable building design, but perhaps they are either slightly misinformed or lacking in the necessary education and experience. At the other end of the scale, there are architects who focus on producing iconic signature buildings; in these projects, our role becomes about working to influence the design to meet legislative requirements while providing a comfortable environment for those occupying the spaces.

Of course that is not to say that sustainable buildings are not iconic, or vice-versa. Heelis, the new headquarters for the National Trust, is instantly recognizable and pushed the boundaries of low-carbon office design. It was a pioneering project that developed a new style of deep-plan, low-rise, naturally ventilated and daylit office building.

Regardless, it is important that architects have an understanding of the role passive design can play in reducing building carbon emissions and providing exemplary internal environments without increased energy requirements. As engineers, we need to be involved as early as possible in the design process, influencing initial design concepts to respond to local site constraints such as noise or overshadowing, and considering predominant winds and solar orientation.

Developing this approach to reducing carbon emissions without adding to the cost is fundamental. As a starting point, we need maximize the use of passive or bioclimatic design. Engineering the form, fabric, structure, and envelope of the building can do the majority of conditioning in its internal environment. This ranges from high levels of insulation and air-tightness to use of natural ventilation, thermal mass, and maximizing daylight.

The structure and envelope are the fundamental elements of any building and should be built right. By working in harmony with the natural environment and the weather patterns driven by free solar energy, we can reduce energy consumption and running costs, and have less mechanical plant to maintain.

However, carefully considered passive design can only condition a building so far. In spaces with high occupancy, intense use, or strict conditions, or in locations with extremes in external temperatures, we will need to supplement with mechanical conditioning to maintain comfort or operational requirements. Well-integrated and efficient active systems—both mechanical and electrical—should work to enhance the passive design. Renewable energy technologies need to be integrated within the building form and engineering systems, rather than being thought of as add-ons. Furthermore, they should be used to offset remaining energy loads after considered passive and active design rather than being installed as the primary means of reducing emissions.

As legislative requirements tighten on allowable carbon emissions, integration of the passive, active, and renewable systems becomes more critical. Architects and engineers need to work seamlessly to craft these into the building design. As legislation is changing, the financial framework that helps to fund renewable technologies is constantly under review. In the United Kingdom in 2010, government grants for the installation of renewables changed from capital funding to payments per kilowatt-hour for generation over twenty to twenty-five years, both for electricity and for renewable heat from sources such as biomass or ground-source heat pumps.

Each time the legislation or funding changes, the goalposts shift, resulting in changes in the balance and emphasis of technologies and potential payback. It is the environmental engineers' role to keep a handle on this, interpret it, and develop the most appropriate strategies for the project. The most successful projects tend to be close collaborations between the design team and an enlightened client willing to explore innovation. This means that communication of these strategies to the rest of the team is key, to get both their acceptance and their valuable input. The diversity of clients, design teams, and budgets means that in practice this communication often involves altering perceptions and managing expectations. Sustainability is a complex topic, and it is useful to try to find tools to help explain it.

The brief for Heelis, the National Trust headquarters, included the request for a building that would be "best practice, occasionally innovative but not pioneering" in sustainability. While there are recognized metrics for measuring the sustainability of a project, such as

LEED, they are often not easily communicated to clients—the detail is embedded within lengthy background documents. Together with Feilden Clegg Bradley Studios architects, we wanted to find a better way to consider their brief and communicate it to the wider team. The result was the "Sustainability Matrix."

The criteria considered in the Matrix were initially the main ones from BREEAM (the UK equivalent to LEED), with additional focus on predicted energy use and carbon emissions. Four columns are identified: minimum standard (building regulations and legislative minimum), best practice, innovative, and pioneering (a zero-carbon building pushing building technologies and construction techniques to their maximum extent). For each of these criteria, the most suitable response to each standard was proposed. In this way, the possibilities for the scheme were explored and the extent of innovation agreed.

It was possible using the Matrix to highlight what could be achieved within budget as well as additional areas that might reasonably improve the overall sustainability of the scheme, but would require additional investment. A payback based on carbon emissions and running cost savings can be calculated for each. In the case of Heelis, the building was paid for by an institutional funder that leased it back to the National Trust on a twenty-five-year lease. Therefore the National Trust directors agreed the design should not include any additional element with a payback of longer than twenty-five years. If the payback was less than twelve years, it was to be included by the design team, and if between twelve and twenty-five years, it was to be taken to the directors for approval. The Matrix provided a useful tool for demonstrating how we could meet or even exceed the brief while involving the client in decisions.

When defining how far a project might go in delivering sustainability, clearly the budget and the client appetite for innovation are key drivers. This becomes further complicated when the procurement route and contractual arrangements are factored in. We must understand where risk is being lodged and who is taking it on. In the case of Heelis, in addition to it being institutionally funded, it was procured under a design and build contract. The design team were novated to the design and build contractor at RIBA Stage D.

The brief had two occasionally conflicting key drivers. The National Trust wanted a building that was clearly the headquarters of a charity with strong environmental values, whereas the institutional funders required an office meeting standard specifications to lower up-front costs and funding requirements. Once this challenge had been met, the design had to be communicated to the design and build contractor so that they could deliver on the brief and design. The key to achieving this was early value engineering and being able to demonstrate the robustness of the design. Whenever an element of the design was challenged by the contractor, it was possible to prove it was integral to the success of the scheme.

Innovation needs to be clearly and carefully explained. It should not be confused with experimentation and the associated risk. The most successful innovations are often about taking well-understood and well-used technologies and applying them in a novel combination or methodology and ensuring that the application is thoroughly considered. Maintaining the same design team from one project to another can spur this approach to innovation and drive the agenda forward. What may have been novel on the first project can be tried and tested by the second, enabling it to be incorporated without substantial discussion. In this way, effort can be focused on pushing the boundaries in other areas to strive toward the target of zero-carbon buildings.

GREEN OFFICES SUSTAINABILITY MATRIX — PAGE 1 OF 2: ENERGY CRITERIA

MAX FORDHAM

Sustainability Criteria		Minimum Standard	Best Practice	Innovative	Pioneering	Notes
Proposed Building Regulations		2010 Part L Regulation	2013 Part L Regulation	2016 Part L Regulation	2019 Part L - 'Zero Carbon'	
1 CO₂ Emission design target		30 kg CO₂/m²/yr	21 kg CO₂/m²/yr	8 kg CO₂/m²/yr	0 kg CO₂/m²/yr 'Carbon Neutral'	'Zero Carbon' not yet fully defined. Typical design stage modelled target
2 DEC rating		C rating	B rating	A rating	A+ rating	Target DEC used rather than EPC - highly user dependent
3 Energy consumption						
	Heating & hot water load	61 kWh/m2/yr	46 kWh/m2/yr	30 kWh/m2/yr	15 kWh/m2/yr	Approximate values. Defined by A) The design Strategy, which is the base installed load and controls strategy defined by the design team; and B) The operation, which is under user control
	Electrical base load	16 kWh/m2/yr	15 kWh/m2/yr	13 kWh/m2/yr	12 kWh/m2/yr	
	IT and small power	48 kWh/m2/yr	41 kWh/m2/yr	33 kWh/m2/yr	26 kWh/m2/yr	
4 On site energy generation		Up to 20% based on local planning	>20% on site renewables	>60%	> 100% on site generation or agreed off-site generation	Highly site specific.
5 U-values (W/m2K)						
	Wall	0.35 (Part L 2010)	0.2	0.15	0.1	Difficult to pass 2010 Building Regs using minimum regulation values. 20%-30% improvement in U-values and airtightness typical.
	Average window	2.2 (Part L 2010)	1.4	1.1	0.8	
	Roof	0.25 (Part L 2010)	0.15	0.12	0.1	
	Ground floor	0.25 (Part L 2010)	0.15	0.12	0.1	
6 Airtightness at 50 Pa		10 m³/h.m² (Part L 2010)	3.5 m³/h.m² (BCO guide)	2 m³/h.m²	1 m³/h.m²	
7 Building occupancy		50-80% Desks occupied at any time of working day.	Hot desking/desk sharing for peripatetic staff. Cleaners/night-security aware of energy use	Hot desking, remote working, 24 hour use restricted to small areas.		Energy use and Carbon emissions could also be considered per person day worked.
8 Controls, metering and monitoring		Seasonal Commissioning. Produce DEC, report to senior management	Commissioning company retained to monitor over first year. Post occupancy evaluation. Action plan to respond to annual DEC	Responsibilities for reading, reviewing, actioning changes defined. Anonymised external reporting. Departmental energy targets	Continual monitoring, fine-tuning and feeding back. Formal external review. Results published to industry. Energy use reward/penalty system	
9 User involvement		Facilities Staff trained at building handover. Building Log Book provided with O&M Manual	Facilities staff involved in commissioning. Non-technical user guide produced and all staff indicted. Energy use fed back to users	Soft landing framework followed (see note). Interactive online user guide. Energy use on interactive display screen and online	Departmental energy use feeds into personal carbon trading (eg WSP's PACT scheme). www.softlandings.org.uk	Often a result of poor commissioning, training & management. www.softlandings.org.uk
10 Summer thermal targets for energy reduction		CIBSE / BCO design targets. Air conditioned Spaces: 24 C +/- 2C. Naturally ventilated: 25°C for <5% and 28°C for <1% working hours. External temperature to suit geographic location	BCO Design Targets used, test the design to UKCIP2020. Dress code partly relaxed in warm weather as ISO7730	Maximise adaptive comfort: internal temperature 2°C < external temperature when external temperature> 27°C. Dress code entirely relaxed. Eg allow shorts and short sleeves in summer. Building design tested to UKCIP 2040	Building design tested to UKCIP 2060	Highly dependent on how staff use the building
11 Thermal mass, ventilation and cooling		Natural ventilation where possible, otherwise mechanical ventilation and comfort cooling. VRV/VRF system used in Server room. Server room set point no less than 24°C	Thermal mass in roof. Natural ventilation plus low grade cooling or mixed-mode with heat recovery. Server room uses free cooling when possible	Natural ventilation with comfort cooling served by GSHP or mech vent with heat recovery. Free cooling and heat recovery to server room	Free cooling = directly coupled cooling	
12 Solar control		Provide fixed external shading. Manual internal blinds	Orient and size windows for capturing useful daylight only. Provide some level of external shading with upgrade strategy to deal with future hotter summers. Solar control glass, mid-pane blinds etc	Automatic adjustable external shading. Consider use of deciduous planting	As innovative plus insulated shutters/blinds with reflective outer coating	
13 Daylighting		Average 2% daylight factor where possible. Views to outside. Glare control blinds	Narrow plan floorplate or rooflights to provide daylight. Views to sky. 80% floor area >2% average daylight and uniformity 0.4	Building form heavily influenced by daylight design. 80% floor area >3% average daylight factor	At least 80% of the floor area has an average daylight factor of 5%. Reflection onto vertical surfaces to reduce perceived gloominess. Building form led by daylight design	
14 Artificial lighting and controls		300-500 lux to BCO and CIBSE guidelines. PIR detectors in WCs etc. Fluorescent fittings throughout	300 lux background lighting plus task lighting. Daylight dimming and presence detection throughout building	150-200 lux background & wall-washing plus task lighting. Daylight dimming. Daylight dimming & presence detection	As innovative with new lighting technologies eg. LED's	Design to CIBSE Lighting Guide 10. BS8206 Part 2 and the BRE Site Layout Guide 10. Design to SLL Lighting Guide LG7
15 IT strategy		Users encouraged to switch off PCs overnight.	Kill switch for non essential peripherals. Servers ramp down under part load. Consider laptops throughout	Thin client system - lower power terminals with centralised computing. Servers running virtualisation software	Off-site internet-based cloud-computing systems	cloud-computing = software and resources provided by internet on demand, like the electricity grid

Row group labels:
- Building and Operational Targets
- User and Operational Interaction
- Design considerations and strategies

Several tools, software programs, concepts, and approaches appear and reappear as teams strive to identify the optimal solution to a given problem. Sometimes the applications of the tools are the goal, other times the tools become just a means to a desired or unexpected outcome. We always direct the class toward the latter.

Computation and Design Optimization

Algorithms in Design: Uses, Limitations, and Development

Lee-Su Huang

The past twenty years have seen rapid advances in the proliferation and application of digital technology to the architectural field, which have profoundly affected the way architecture is conceived and constructed. Traditionally, the essence of an unbuilt building has always been explored through the construction of physical models and drawings, conveyed to the contractor through 2D representational techniques such as plans, sections, and detail drawings.

New digital design technologies have made it possible to conceive a building from conceptual design development to construction management almost entirely within the virtual space of the computer. Within the realm of digital design, algorithms represent one of the forefronts being currently explored due to their potential application in the design and construction environment. The use and purpose of algorithms is being hotly contested within the design profession; questions have been raised about the authorship and authenticity of designs produced by a process-based program. Ultimately, algorithms are tools, logic-function processes that can be applied to simplify complex tasks, automate massively mundane and repetitive operations, and explore possibilities unattainable (for all practical purposes) within the confines of human cognition. This essay will discuss some of the possible applications of algorithms in architecture, citing examples that have been carried out or are under development, and also projecting into the future possibilities of its application within the architectural design and construction context.

Currently there are two leading trends in the use of algorithms within the architectural context: one deals with the overall strategies of schematic design in its many disparate forms, including generative form-finding processes, optimizations responding to certain criteria, and emergent design behavioral systems that explore the use of algorithms in the larger context of the scheme as a strategic stance. The second major trend is the use of algorithms in the production phase of a project, as in the automation of hugely repetitive tasks to increase efficiency and precision, or the translation of a predetermined schematic design into detailed fabrication documents for use in the digital fabrication production chain.

In parallel, construction automation and digital fabrication are also fields that have seen substantial development in the past twenty years. Buildings such as the Bilbao Guggenheim by Frank Gehry demonstrate the mass-customization abilities of modern construction and fabrication technologies. However, as digitally driven forms and design explorations gain complexity, algorithms often present the only solution to dealing with the highly irregular geometries present in such projects. The technical aspects and digitally driven processes that are practically impossible with manual human labor is what make the architecture truly digital, and algorithmic processes essential to its existence.

Beijing National Aquatics Center.
PTW Architects, CSCEC, CCDI, Arup,
2004-2007

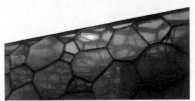

Façade detail

Case One:
Beijing National Aquatics Center

The first example is the Beijing National Aquatics Center, also known as the Water Cube. A collaboration among the China State Construction and Engineering Company, Australia-based PTW Architects, and Ove Arup Structural Engineers, the Water Cube is a space frame covered with ethylenetetraflouroethylene panels configured as an array of polyhedra. The solution uses a configuration called the Weaire-Phelan structure, which mixes two cells of equal volume, an irregular pentagonal dodecahedron and an irregular tertrakaidecahedron, truncated at an arbitrary angle to convey an organic and random sense to the structure despite the overall array being totally regular. The discovery of the Weaire-Phelan structure or "foam" configuration in 1993 by Trinity College Dublin physicist Denis Weaire and his student Robert Phelan was the result of running mathematical simulation software to find an optimal solution to the problem of partitioning space into equal-volume cells using the least interface area, or in other words what is the most efficient bubble foam?

Translating this highly irregular structure into something buildable is a challenge best tackled with algorithmic processes. Using the wireframe model of the structural centerlines, a Visual Basic routine in Microstation was created to automate the process of translating the wireframe model into a steel structure model, bridging the link between the engineering and analysis model to the working 3D CAD model. With a total of 22,000 steel tubes, 12,000 nodes, and 4,000 different cladding panels, the project would have been practically impossible to produce with manual labor within the given development timeframe. By the conclusion of the project, the production process had been largely automated through scripting. One program generates the entire

Construction photo

geometry from scratch, based on the characteristics of the Weaire-Phelan foam and the building envelope, while a structural optimization routine was written to assign member sizes to all steelwork and joints, and another to translate the structural analysis model into a 3D CAD model suitable for outputting construction drawings automatically.

Looking at this particular example and other similar schemes that apply algorithmic geometric principles as integral parts of the building fabric (such as the Serpentine Pavilion by Toyo Ito and Cecil Balmond, and the Beijing National Stadium by Herzog and de Meuron), an observation can be made regarding the overall characteristics of the designs: they all exhibit an homogeneity that, at the very least, facilitates the optimal application of algorithms to generate the bulk of the building's structural framework with minimal manual intervention. This purity of the architectural scheme inherent to the building typology makes possible the universal application of algorithms, while other more site-specific and function-specific building typologies would have required much more manual fine-tuning to allow the algorithms to respond to local conditions. It is apparent here that algorithms are suited to run through iterative processes of similar overall schemes, but are not as suited to adapting to localized variations where input parameters are so disparate that the expected reasonable outputs cannot be generated, such that the algorithm "breaks."

[C]space DRL Ten Pavilion. Alan Dempsey,
Alvin Huang, Adams Kara Taylor, 2007-2008

Case Two:

[C]space DRL Ten Pavilion

The second case is the [C]space DRL Ten Pavilion, designed by Alan Dempsey and Alvin Huang and developed in collaboration with structural consultants Adams Kara Taylor and the Austrian concrete firm Reider and Company. A prime example of pioneering the workflow process from digital to physical, the pavilion is conceived as a discontinuous shell structure made of 13-milimeter-thick fiber-reinforced concrete sheets cut to 850 individually different profiles, with 2,400 joints. The project is remarkably akin to a full-scale version of a paper model built with rib profiles and notch joints. Due to the irregular geometry and resulting intersection angles of the profiles, a number of scripts were written throughout development of the project to deal with the vast amount of information.

Soliciting help from Marc Fornes of THEVERY-MANY to aid in Rhinoscripting, code was written that generates a pattern for the segmentation of a surface based on subdividing its isocurve. Another script was used to unroll, flatten, nest, and annotate the panels for fabrication. Later iterations of this script were applied to notch the intersection joints into the profiles, and additionally automatically adjust notch widths for construction-tolerance purposes.

This project demonstrates how in certain circumstances algorithms must be adapted to deal with local geometries and operations. This exposes some of the limitations of algorithms when used in this capacity: their input/output parameter precision exacerbates the need to tailor the code to project circumstances. This negates the ability for some of these algorithms to be used as a more general tool. As usage of algorithms proliferate in both the

construction and design professions, however the possibility exists of creating an open-source library of basic algorithms for adaptation, allowing designers to mix and match, or construct more complex routines out of several basic algorithmic operations.

Pushing this concept of open-source and user-customizable execution modules is McNeel's Grasshopper plug-in for Rhinoceros. The functional snippets of code are packaged into "components" that can form the building blocks of larger, more complex execution sets and rearranged easily by the user to adapt to different requirements or conditions. The ease of use of this "graphical scripting interface" has given Grasshopper significant traction within the industry in the past few years. If the user has scripting/programming capability, custom modules can be created if there are no equivalent combinations within the provided components. The package offers several advantages to the average architectural designer: ease of use, flexibility and adaptability, dynamic visual feedback, close integration with the modeling package, and open-source attitude on the part of the developer. This can be seen from the countless user-created components and plug-ins shared on the internet, like the live physics engine plug-in Kangaroo, or the topological transformation mesh plug-in Weaverbird. Other plug-ins serve as interfaces between Grasshopper and full-fledged programs such as Gecko, which interfaces with the energy analysis package Ecotect, or Firefly, which connects to the Arduino microcontroller for interactive installations. On the other hand, the mono-directional execution nature of the program makes the construction of logical evaluation loops more difficult, which is something the pure scripting languages are very well suited for.

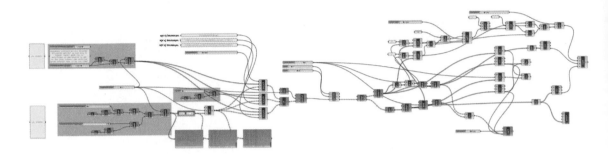

Grasshopper definition for babel-On. Lee-Su Huang,
Gregory Thomas Spaw/SHo Architects, 2010

Case Three:
MVRDV and Functionmixer

Let us turn our focus to the more strategic and speculative applications of algorithms in design. The Dutch firm MVRDV has pioneered the use of computational programs in strategic planning and urban design since 2000, with their development of programs such as The Functionmixer and The Regionmaker. Urban planning is a profession that requires the interpretation, integration, manipulation, and representation of vast amounts of data. This evaluation process is ideally suited for exploration through usage of algorithmic processes. The Functionmixer provides a framework for the functional optimization of a mixed-use neighborhood. Using a weighted stochastic search algorithm, the software runs through millions of possible zoning combinations while analyzing and ranking each, ultimately suggesting an optimal combination according to a given input of desires. Conceptually, this kind of "brute-force" approach is possible only through computer simulation and also holds the potential to uncover configurations inconceivable through normal rational human thought processes. As a tool, this program provides insight into the effect and limits of current laws, while serving as a vehicle

for comparing possible changes and solutions. The Regionmaker uses the same core concept as The Functionmixer, but adapted and applied to a much larger scale, while integrating and responding to existing criteria such as regional demographics and GIS data on roads, infrastructure, energy requirements, and so forth. Within this program, varying regional configurations can be tested and evaluated to analyze the limitations of each system and identify where potential problems may lie.

Although these hyper-rationalist models of urban planning may never exist in reality, the quantitative representation of urban forces is a useful tool in understanding the dynamics of the overall systems, while also serving as a platform in which hypotheses may be tested to see the effects chronologically. As the simulation grows more complex with more parameters being modeled, it will become more accurate and useful. The underlying algorithms functioning as mathematical constructs to describe real-world conditions can be updated or refined as urban theories evolve. Such relatively closed theoretical systems are particularly suited for the application of algorithms.

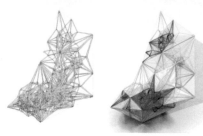

Super Tall: Performance and Atmosphere.
Gregory Thomas Spaw, 2007

British Museum Great Court,
Foster + Partners, 1994-2000

Beijing National Stadium, Herzog & de Meuron,
ArupSport, CADG, 2003-2008

Case Four:
Algorithmic Architecture

On the slightly smaller scale of the individual building, quite a number of schools have been exploring the possibilities of applying algorithms intelligently in the context of architectural design. In the course titled Algorithmic Architecture at the GSD taught by Kostas Terzidis, students use Maya's MEL scripting language to explore formal and performative associations in architecture with algorithmic methods. Using algorithmic processes such as shape grammars, cellular automata, genetic algorithms, stochastic processes, L-systems, and Marcov chains, the behaviors of such systems and their limits are tested and adapted—sometimes as purely formal explorations, sometimes as applied to concrete functional space.

Additionally, there has been a burgeoning of architectural solutions based on a mathematical premise, be it for façade patterning purposes or to describe the curvature of a complex canopy. This current interest in the spatial implications/applications of mathematical issues has led to projects based on space packing formulas, aperiodic tiling, minimal surfaces, and magnetic field vectors. These range from large projects that have been built, like Norman Foster's Great Court canopy at the British Museum, to small-scale fabrication or installation projects such as the Minimal Complexity installation by Vlad Tenu at the University of Houston. There are an increasing number of projects that engage algorithmic processes from the inception of the project all the way through to its production. More often than not, the choice of the mathematical basis is made with consideration of the structural or fabrication characteristics, providing a synergy between the conceptual mathematical generation of the project, the structural performance, and fabrication techniques.

Such exercises provide insight into the tendencies and underlying logic structure of these processes and their inherent range of intelligence, randomness, and unpredictability. With further development, any of the given systems could be evolved into a tectonic architectural strategy with practical real-life application. Similar explorations and programs are in place at leading institutions around the world, and also within cutting-edge practices well versed in the computational aspects of design.

Beijing National Stadium, Herzog & de Meuron, ArupSport, CADG, 2003-2008

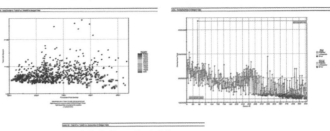

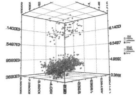

ModeFrontier statistical analysis chart. Lee-Su Huang, Stubborn Urbanism 2009

Case Five: Genetic Algorithms

The next generations of algorithms have begun to include evaluative processes within their framework, allowing them to evaluate and adapt to conditions within certain parameter sets, giving them "smart" behavior characteristics. This makes them more adaptive to a wider range of inputs, and when given the right conditions, they can exhibit a "fuzzy logic."Along this line of reasoning in teaching the algorithms to "choose" a more desirable result are genetic algorithms. This class of algorithms emulates the genetic evolution and mutation process, combined with fitness criteria and selection feedback loops built in. This allows the algorithm to truly evolve the solution over time as it runs through iterations to learn, grow, and adapt, while converging toward possible optimal solutions. They are especially suited for multi-objective optimization of conflicting goals such as material strength, weight, and thickness. Widely used in the engineering world to design and analyze high-performance machines such as Formula One cars and yachts, the application of these algorithms has just begun to take hold in the architecture field. One recent high-profile example is the Beijing National Stadium by Herzog and de Meuron. While the primary structural members are arrayed in a uniform radial configuration, a second layer of interweaving bracing elements form an irregular web pattern that was optimized to have a uniform opening distribution, minimizing localized stresses on the structural framework.

David Benjamin has been conducting this type of design research through his studios at Columbia GSAPP for several years now, utilizing a hybrid workflow of CATIA, Esteco's modeFrontier, and the Autodesk ROBOT structural analysis package. More recently, David Rutten of McNeel software has developed the Galapagos plug-in for Grasshopper, an evolutionary solver that brings the capabilities of genetic algorithms to the Rhinoceros/Grasshopper software ecosystem. While still comparatively rudimentary compared to the software packages in use by the engineering industries, it points toward a new direction that has enormous untapped potential. Two major things are worth noting about this approach. First, the optimization may or may not converge on one solution within the design space. There may be a number of equally optimized solutions that are differently weighted toward certain fitness criteria, forming a so-called Pareto frontier. It is then up to the designer to exercise both design and aesthetic judgment to choose among equally valid solutions. Second, due to the brute-force nature of the strategy in examining hundreds and thousands of possible solutions, the method is computationally expensive, and more complex problems will take exponentially more time to solve. Although the speed of computer hardware has increased greatly in recent years, this is still a factor to take into account.

Conclusion

The current use of algorithms has focused primarily on top-down strategic investigations within relatively closed systems, where it is used to quickly and efficiently translate design models into construction or fabrication documents. While building information modeling (BIM) packages such as Autodesk Revit have begun to solve the problem of design documentation and management, its design exploration tools are still fairly limited, making it more suited for projects with many repetitive elements. What of the middle spectrum?

With parametric modeling software such as Dassault's CATIA and Gehry Technologies' Digital Project gradually making inroads to the design profession, designers are being exposed to tools previously unavailable. Coupled with software capable of finite element and computational fluid dynamics analyses, environmental analysis suites such as Ecotect and Radiance, designers can predict and simulate the structural and performative behavior of buildings more accurately than ever before. However, as a software package with industrial and engineering roots, CATIA remains a somewhat inaccessible program for those not exclusively trained in its operation, and its cost of ownership remains high except for large projects that provide economies of scale.

Bridging the gap slightly more successfully is the Grasshopper parametric environment, which approaches the interface and usage model more from a designer's point of view. There are still problems with its relative inability to deal with recursive design logics, which is somewhat mitigated by the ability to write custom components. Several structural analysis and BIM plug-in packages are in development, which will produce a more rounded solution for the overall design process.

In time, a more unified, cohesive approach will be developed to merging parametric and BIM modeling with algorithmic scripting solutions. The streamlining of the design process from conception to construction as part of the same workflow, with little to no disjunction, will open endless possibilities for the profession, while making integration between disciplines much easier.

Bibliography
Maas, Winy. *Five Minutes City* (Rotterdam: Episode Publishers, 2003).
Terzidis, Kostas. *Algorithmic Architecture* (Oxford: Architectural Press, Elsevier, 2006).

Integrated Design: A Computational Approach to the Structural and Architectural Design of Diagrid Structures

Jessica Sundberg Zofchak

For most of my life, I've defined my interests somewhere between the creative, imaginative field of architecture and the more calculated, systematic approach of engineering. While studying design, I've strived to validate my design decisions with substantial concepts, however scientific. I see architectural design as an opportunity to intelligently design a product–by taking structural opportunity cues, environmental cues, lighting directions, climate needs, cost, and end-user desires into consideration. Building design becomes an integration of architecture with engineering, physics, mathematics, and product development and marketing. The desire to consider all disciplines in the design makes for a process that, if left to traditional methods, is disorganized and nearly impossible to coordinate.

When expanding this perspective to the design of actual buildings, coordination is even more necessary. Not ten years ago, firm-to-firm communication was a complicated network, with project data being released in multiple directions in a variety of formats and data types. As design projects become more complex and sophisticated, this coordination becomes increasingly complicated, yet all the more critical to success.

As technology has progressed, computational methods have become more advanced to deal with building information. Enhanced 3D modeling software has allowed firms to capture geometries never before translated from complex abstract ideas to tangible models. Designs once limited to conceptual endeavors now reach closer to becoming reality. With modeling becoming more sophisticated, it is now the coordination of building information among disciplines that limits the range of design.

Intent

The following scripting implementation attempts to take a structural system of particular interest and, starting at the initial architectural concept design level, iteratively design the structural system within available software. Using coding tools such as Rhino Script and Visual Basic, the preliminary framework for a more automated system was created. The limitations of this particular model were assessed, along with the limitations of the computing concept itself. A preliminary presentation of an iterative design model was given to a panel of architects, software engineers, and artists. The feedback on the initial intent–to create a piece of software that takes an input of form and outputs a structure–was surprising.

The architects wanted a tool that they could learn from; they did not want such a "black box." It was for this reason that the more iterative, design-oriented decision-making loop was created.

This approach is meant to be a rough model of what is possible with further software development to integrate architectural design and structural design software. I propose a computational method devised for the development of free-form architectural designs with a structural diagrid system and critique its relevance to the world of design today. The method presented is rooted in an investigation of the effectiveness of technology use for certain projects in decades past and present.

Hearst Tower, New York, NY

30 St. Mary Axe Tower, London

Diagrid Development

The study began as in inquiry into how to approach the structural design of diagrid systems for buildings with irregular geometry, or more free-form buildings.[1] Diagrids or exodiagonal systems are perimeter structural configurations characterized by a narrow grid of diagonal members that are involved both in gravity and lateral load resistance. Many buildings, including tall structures, have been built using the diagrid system. An earlier example of diagrid systems in medium-rise buildings can be found in the IBM Pittsburgh building, "perhaps one of the first uses of diagrid."[2] Norman Foster's Hearst Tower in New York City and his 30 St. Mary Axe (the Swiss Re Tower) in London utilize the diagrid structure for different geometries.

Similarly, Kohn Pedersen Fox Associates' design for the World Product Center in New York City and Skidmore Owings and Merrill's Lotte Tower design for South Korea demonstrate other applications of a diagrid structural system. In general, the notable characteristic of these buildings is their relatively simple geometry. Each building varies little in section and plan, which is either square or circular.

With the trend of architectural concept designs taking on more blob-like or fluid forms, the diagrid structure becomes a strong candidate for the selection out of a range of structural systems. Likewise, it can also be used to more accurately approximate varying surface geometries, such as twisted and tilted forms. To investigate the impacts of various geometric configurations of complex-shaped buildings, such as the degree of fluctuation of freedom, the rate of twisting, and the angle of tilting,

parametric models need to be generated using today's CAD/CAM technology. The results obtained are consequently exported to structural engineering software for design and analyses.

Constructability is a serious issue in diagrid structures because the nodes are more complicated than those of orthogonal structures and tend to be more expensive, leading to a "limiting factor" of the diagrid. Prefabrication of the nodal elements is essential, and pin connections using bolts can be easily constructed at the job site.

Structural and Architectural Advantages

In a diagrid structural system, most vertical and lateral loads of the building are taken by the exterior diagrid elements, with less reliance on the core. This allows for column-free floor plans that are ideal in tower design to provide the optimal rentable space. The greatest advantage of the diagrid structure is its angled elements. These allow it to take both the vertical and lateral loads to improve stiffness of the structure. A stiffness-based approach to design is most fitting when considering a diagrid structure.

A diagrid structure is modeled as a vertical cantilever beam on the ground and subdivided longitudinally into modules according to the repetitive diagrid pattern. Each module is defined by a single diagrid level that extends over multiple stories.[3] The diagonal members are assumed to be pin-ended and therefore resist the transverse shear and the overturning moment (produced by the lateral and vertical loads) through axial action only. In considering a stiffness-based design, the stiffness of the structure can be calculated based on the angle of the module.[4]

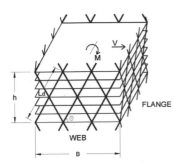

Typical diagrid module

201 Bishopsgate and The Broadgate Towers, London

In a regular diagrid system, the stiffness would be defined as the following in the horizontal direction:

$$K_V \cong \frac{AE}{h}\sin^3\theta$$

Likewise, the stiffness in the vertical direction is the following:

$$K_H \cong \frac{AE}{h}\sin\theta\cos^2\theta$$

Comparing these values to the stiffness of the equivalent rigid frame or braced frame structure (such as Skidmore Owings and Merrill's 201 Bishopsgate and The Broadgate Tower in London), the diagrid provides relatively the same stiffness as the braced frame structure while eliminating the vertical members and thus reducing the amount of material.

Table 1. Stiffness comparison for structural systems

System	Lateral Stiffness
Rigid Frame, Pinned	$K \cong \frac{12EI}{h^3}$
Braced Frame	$K \cong \frac{AE}{h}$
Diagrid	$K \cong \frac{AE}{h}$

Consequently, a performance-based design approach would be to determine the maximum allow-

able displacement at the top of the tower under a specified loading, determine the effective stiffness of the tower to satisfy the displacement limits, and size the diagonals and optimize the module angle θ accordingly to provide the stiffness required by code. Optimal stiffness-based design corresponds to a uniform deformation under the design loading and is possible because the building structure is modeled as a vertical cantilever beam on the ground.

Choosing a diagrid structure has significant architectural implications. While eliminating material, the exclusion of the vertical members allows for a more varied geometric pattern for the structure, as well as fewer obstructions along the façade of the building. The diagrid, as a consequence, can be used to fit irregular surfaces.

Automation Methods
By introducing complex geometries into the diagrid problem, what would already be a challenging design becomes increasingly difficult. With hundreds of members and thousands of different angles, proper design requires more aggressive computational methods. For example, how would a structural engineer begin to design the diagrid of a more free-form tower if the architectural designer supplied only a surface describing the concept's geometry? Or how can all the unique structural details such as connections and joints be defined and fabricated? This automation study tries to solve this issue.
The final geometry of the diagrid is of utmost concern to the designer, as it defines a main visual feature of the building. Therefore the designer has many options to consider. To expedite the design process, it would be ideal for the 3D modeling soft-

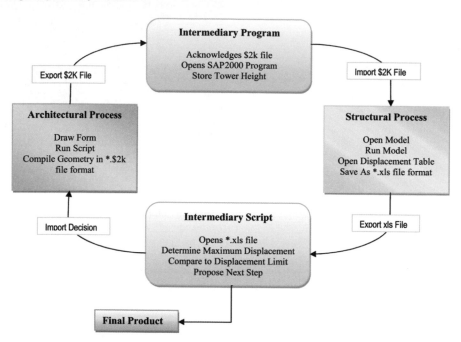

ware to recognize surface geometry defined by the architect and, in real time, calculate whether the desired diagrid geometry is adequate for the building—bringing the role of structural engineer to the forefront of the architectural design process. By using computer coding, scripts can be written that simulate this ideal real-time design iteration process. The flowchart shows the steps that make up the semi-automated process that was developed.

Within Rhinoceros 4.0, the designer, or user, can generate a surface using lofted curves. Then he or she can run a script that will interpret that surface and apply a diagrid based on user-defined inputs. Next, the data indexing the diagrid elements can be compiled and inserted into a file that models the basic structural information necessary for the structural analysis program chosen, SAP2000. Once this data is assembled, the structural analysis can be run, and the exported data regarding the resulting displacement under the given loading cases can be analyzed. If the structural system is sufficient for the tower, the cycle can end; however, if the tower's displacement was not within the required limits of around L/300

to L/500 (when L equals the tower height), the user can make changes that will improve the design, such as upper stories with smaller angle and lower stories with larger angle (gradually varying angles). A module angle equal to 35 degrees ensures the maximum shear rigidity while larger angles ensure better bending rigidity.[5] A balance between these two conflicting requirements should be sought by defining the optimal angle of the diagrid module.

User Input
The designer must input the basic geometry of the tower under investigation along with inputs that define the parameters of the diagrid. Although the curve inputs require greater study and investigation to arrive at a desirable design, the other inputs are simple parameters. The purpose of the program is to allow the user to easily vary these parameters and quickly redraw multiple variations before moving on to export the final architectural data. The table summarizes the necessary inputs for the script along with their limitations or restrictions. See the appendix for images of a design example carried through the steps of the architectural process.

Displacement outcome options

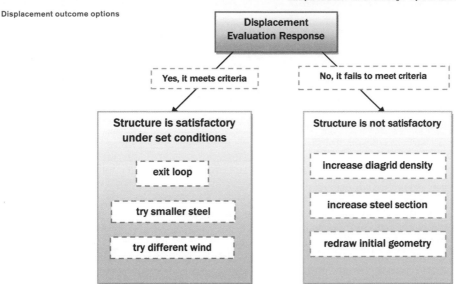

Table 2. User input for script studied

Input	Variable Name	Description	Limitations
one or more closed curves	strGeo, becomes strShape	Model; Used as input to generate the tower geometry by lofting the curves, a Rhino function	Curves must all be in same direction (clockwise or counterclockwise)
number of floors per diagrid module	intGridZ	The number of building stories that the diagrid will span vertically	Division must work out to be a close factor of the total height of the building
number of horizontal diagrid modules	intGrid	The number of diagrid modules that will surround the building	None, but limited by aesthetics and eventual structural considerations. If too many, diagrid system less effective laterally.
wind angle	dblWind	Angle between 0 and 360 degrees with 90 degrees representing due North and 270 representing due South	Angle Input should be between 0 and 360 degrees, otherwise program will still convert to radians

Output and Optimization Results

Once the architectural output data is manipulated within Microsoft Excel, the new file representing the tower can be imported into the structural analysis package for simulation. After the model is run, the program will export the displacement data, and the intermediary software will compare the displacements to the allowable threshold. Then the user faces the decisions outlined in the flowchart. Based on the user's choice, the whole process can begin a second iteration, and so on, until a more refined design is achieved.

Limitations and Relevance

The semi-automated design process that was created has many limitations. The integrated architectural script, however powerful and noteworthy, still has significant constraints. The curved geometry cannot be too extreme (no bubbles or extreme protrusions), there is no embedded variability for the floor-to-floor height, and the modular arithmetic methods used in the drawing of the system crops the height of the original design if peaks or inconsistencies occur at the top of the tower. The script also generates only regularized designs, with no variation of height in the vertical direction. Also, the structural design assumes that the architecture is in fact a tower with a base width to height ratio of at most 1:7.

The first two limitations place unnecessary restrictions on the design of the diagrid system. For example, the diagrid need not have a totally regularized design. Some designers have created concepts for towers that use hybridized braced frame/diagrid structural systems with diagonals at irregular patterns spanning different floor intervals. Moreover, the script in the process above could be modified to randomize the diagrid to provide different triangulation patterns within parameter limits for steel lengths (to avoid buckling issues). As mentioned before, constructability is a serious issue, and a prefabrication strategy must be studied to eliminate its "disadvantage" of complicated nodes on the design process.

Finally, this scripting study investigates only one type of structural system. To test multiple systems for the same tower design, additional scripts must be written that define the elements of such systems. The architect would have only the specific tools that have already been created for a particular structural system.

Looking to the Future
After completing this study, I realized just how difficult it is to integrate structural analysis in real time with an architectural design package that realistically can benefit both disciplines. Yet in time it may be beneficial to create more advanced software packages that consider parameters other than just structural systems. Although the idea of linking structural design and architectural design through embedded scripted calculations is extremely powerful, the array of new tools, both available and possible, has the potential to limit design. When critiquing the scripted integrated design approach, the more overarching design implications must be considered.

Integration with Other Coded Design Methods
In addition to enhancing the coordination with construction processes for fabrication, integrated computer models also have the ability to access other building-specific data to optimize the architecture along many performance measures. Eleftheria Fasoulaki discusses the incorporation of lighting optimization in generative building models.[6] She used a similar design algorithm between RhinoScript,

data storage in Excel, and Ecotect, a lighting and climate simulation program, to generate forms based on inputs of site data and building geometry. She presents a model that lets the user weigh preferences for variables of basic structural, solar, and zoning definitions to give different resulting modes of optimization.

Similarly, the model outlined here for a more detailed structural design could be incorporated with other simulation data to optimize along multiple parameters. These parameters can be defined from real estate development, environmental or mechanical engineering, or urban planning perspectives. At a base level, the parameters can be evaluated to guarantee that fundamental limits have not been met. Then weights can be assigned for other parameters such as daylighting, natural airflow, material use, site interaction or views, and/or seismic performance. For example, the diagrid module spacing can be optimized in two additional ways: 1) for use of predetermined member lengths rather than strict user input for spacing, allowing for more regular sizing that could reduce material waste or simplify the fabrication process, and 2) to allow for unobstructed views on certain levels of the building in set directions.

While building information is assembled through BIM modeling, firms typically use in-house simulation and design software to generate the systems that are then drawn within shared models. If companies from all disciplines of building design moved away from the use of proprietary software and toward the sharing of coordinated data and modeling (simulation/design/testing rather than geometry models), buildings could be optimized along multiple parameters from the start of the design process. Architects could then gain an intuition for the contribution of different requirements on building design. If a building is designed more intelligently, time and effort can be saved throughout the remainder of the project. However, one must not delegate too much authority to the tools of design.

Limitations of Full or Completely Integrated Design
The natural progression of architectural design flows through a problem statement, conceptual design

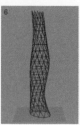

Steps 1 to 7

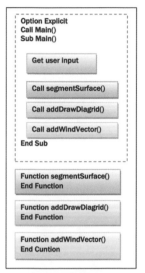

```
Option Explicit
Call Main()
Sub Main()

    Get user input

    Call segmentSurface()

    Call addDrawDiagrid()

    Call addWindVector()

End Sub


Function segmentSurface()
End Function

Function addDrawDiagrid()
End Function

Function addWindVector()
End Cuntion
```

Pseudocode for DrawDiagrid function

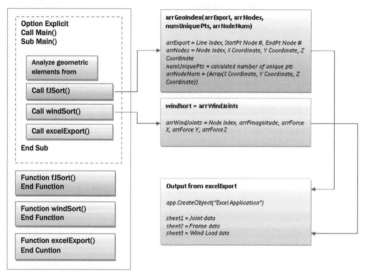

```
Option Explicit
Call Main()
Sub Main()

    Analyze geometric
    elements from

    Call fJSort()

    Call windSort()

    Call excelExport()

End Sub


Function fJSort()
End Function

Function windSort()
End Function

Function excelExport()
End Cuntion
```

Pseudocode for geometric and load data export

arrGeoIndex(arrExport, arrNodes, numUniquePts, arrNodeNum)

arrExport = Line Index, StartPt Node #, EndPt Node #
arrNodes = Node Index, X Coordinate, Y Coordinate, Z Coordinate
numUniquePts = calculated number of unique pts
arrNodeNum = (Array(X Coordinate, Y Coordinate, Z Coordinate))

windsort = arrWindJoints

arrWindJoints = Node Index, arrFmagnitude, arrForce X, arrForce Y, arrForce Z

Output from excelExport

app.CreateObject("Excel.Application")

sheet1 = Joint data
sheet2 = Frame data
sheet3 = Wind Load data

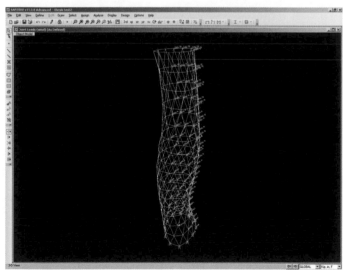

Imported model to SAP2000

	A	B	C	D	E	F	G	H	I	J
1	TABLE: Joint Displacements									
2	Joint	OutputCase	CaseType	U1	U2	U3	R1	R2	R3	
3	Text	Text	Text	m	m	m	Radians	Radians	Radians	
4	1	ACASE1	LinStatic	1.794789	0.104156	-0.024513	-0.000558	0.015306	0.00783	
5	2	ACASE1	LinStatic	1.790354	0.087938	0.006573	-0.000634	0.015255	0.008015	
6	3	ACASE1	LinStatic	1.796567	0.071635	0.037629	-0.000657	0.015182	0.008187	
7	4	ACASE1	LinStatic	1.811211	0.062197	0.056268	-0.000637	0.015248	0.008157	
8	5	ACASE1	LinStatic	1.828089	0.065342	0.051687	-0.000692	0.015393	0.007952	
9	6	ACASE1	LinStatic	1.838217	0.078138	0.027505	-0.000694	0.015412	0.007828	
10	7	ACASE1	LinStatic	1.837684	0.094789	-0.004975	-0.000541	0.015348	0.008094	
11	8	ACASE1	LinStatic	1.826137	0.10822	-0.030624	-0.00053	0.015285	0.008294	
12	9	ACASE1	LinStatic	1.809196	0.11246	-0.039654	-0.000553	0.01529	0.007991	
13	10	ACASE1	LinStatic	1.775172	0.098107	-0.015182	-0.000499	0.015307	0.008302	
14	11	ACASE1	LinStatic	1.76031	0.109898	-0.038112	-0.000538	0.015371	0.007993	
15	12	ACASE1	LinStatic	1.784205	0.081417	0.016522	-0.000613	0.015342	0.00806	
16	13	ACASE1	LinStatic	1.839005	0.078092	0.027666	-0.000695	0.015412	0.007829	
17	14	ACASE1	LinStatic	1.778483	0.064791	0.048343	-0.000709	0.015415	0.007868	
18	15	ACASE1	LinStatic	1.762667	0.055937	0.064198	-0.000671	0.015338	0.00797	
19	16	ACASE1	LinStatic	1.745049	0.060488	0.054133	-0.000666	0.015204	0.008126	
20	17	ACASE1	LinStatic	1.735256	0.076518	0.023488	-0.00067	0.015184	0.008164	

Table 3. Displacement output

approaches, iteration and refinement, and finally, a proposed design. The forms available to an architect are nearly limitless, while the chosen form may have specific, prescribed structural needs. Even though the overall structural typology may be decided, the intricate design of the structure may have a variety of solutions. It is here that the largest issues with embedded scripting begin to unfold.

A script, as such outlined here, was written to perform a specific algorithm. That algorithm may have a variety of conditional networks that compare variables and make decisions based on those variables, yet all possible outcomes must be inherent in the script. Solutions that a computer arrives at may be unusual and therefore subject to evaluation and possible rejection through human judgment; however, a solution can never be entirely born from the computing tools. The computer cannot generate something that it has not been taught to create. Therefore the tools available to a designer can create only that which has already been designed or coded. The computer will not hybridize or create new structural systems without being told how to do so via a

genetic algorithm with parameters and associated weights or values defined by the user or a genetic optimization method.

Finally, the full integration of structural design into architectural design tools raises an important issue of liability and safety responsibility. Inserting intelligent structural design tools into architectural design cannot take the place of the structural engineer. In the future, the role of the engineer would not be minimized, but rather, more present in the design development. Engineers must help define preliminary limitations for the architect along with sets of parameters for building elements. A designer can improve his or her work by having knowledge of constraints.

Design Implications

This has serious implications for the prospects of innovative engineering and design. It seems as though we are in danger of taking an innovative tool and destroying innovation itself. If these tools are relied on completely, we limit ourselves as designers—both structural and architectural—to the cards we hold. Design no longer evolves through discoveries and lessons learned on a project-to-project basis. Lessons are no longer carried over to act as a framework and foundation for subsequent design challenges. With these tools, we limit ourselves to the solutions that have been devised before, unless we innovate and expand the code to allow for the new methods.

Integrated Design Conclusions

By trying to coordinate the interaction between structural engineer and architectural designer earlier in the design process through predefined computational methods, the creative design process can suffer. With coordinated data and the inclusion of innovative minds, however, the design process can thrive. A balance must be reached between drawing methods and computational solutions to not place restrictions on the design too early in the process. The coordination of building models can be beneficial to architects and engineers during the long months or years of collaboration. When approaching each project, the design team should define its own requirements for technology. When used wisely, computational methods can be an

asset to a project's development, aiding in data coordination (compatibility between disciplines), concept modeling for client approval (Henderson Waves Bridge), façade analysis (John Lewis Department Store), drawing-set production (possible use: Phaeno Science Center), fabrication of building elements (possible use: Phaeno Science Center, Henderson Waves Bridge, Queen Mary façades), and construction.[7]

The advancement of technology must be welcomed, but applied with caution on a project-to-project basis. The computational methods used must not constitute the only collaboration in design between the architect and engineer. Software cannot replace the relationship between team members of both disciplines, yet it can be used to enhance collaboration and coordination of design goals.

Appendix

The images included here show the algorithmic steps performed through a design example—from architectural design through structural analysis and iteration/optimization.[8]

Notes
1. The following method for the integration of design and engineering for tall towers with diagrid structures was developed for completion of the thesis requirements of MIT's High Performance Structures program in 2009. What started as a deviation from a required assignment for the School of Architecture and Planning's IAP 4.195 Rhinoscripting: Control and Command class soon became the foundation of a thesis. It seemed like an exploration into new territory, but it was soon revealed to me that certain design-oriented structural firms such as AKT and Buro Happold were already pursuing the development of similar methods.
2. R. G. Weingardt. "Skyscraper Superstar – Leslie Earl Robertson", *Structure Magazine* (June 2007) pp. 60-64.
3. K. Moon, J. J. Connor and J. E Fernandez. "Diagrid Structural

Systems for Tall Buildings: Characteristics and Methodology for Preliminary Design", in *The Structural Design of Tall and Special Buildings* (Vol. 16.2, 2007), pp. 205-230.
4. ibid.
5. R. Arpino. "Diagrid Structures for Tall Buildings," Thesis (in Italian), 2010. Department of Structural Engineering, University of Naples/Italy.
6. Eleftheria Fasoulaki. "Integrated Design: A Generative Multi-Performative Design Approach," MIT thesis, June 2008.
7. Hanif Kara and Andreas Georgoulias. Harvard University GSD 6328: "In Search of Design Through Engineers," Spring 2009.
8. Jessica Sundberg. "A Computational Approach to the Design of Free Form Diagrid Structures," MIT thesis, May 2009.

Form Finding: The Engineer's Approach

Fernando Pereira Mosqueira

Hotel Arts Barcelona, Firth of Forth Bridge, San Siro Stadium. (Clockwise from left)

Optimization tools have suffered from a lack of a development during past decades. Before computers, small problems were resolved analytically, and these studies led to research on methods to optimize any given function. Among the methods, conjugate gradient method (CGM) and line search (LS) were proven to be the most efficient for structural optimization. Several software companies, such as Optistruct (Altair Hyperworks), implemented this type of tool, but the high price and the substantial knowledge required to use it limited the software to high-tech engineers. Furthermore, the complexity of this type of software does not meet the requirements of typical users, who just want to catch a glimpse of what kind of geometry and sizing would be optimum given a set of constraints. Here I try to explain a way to accomplish structural shape and size optimization at the same time in the MATLAB environment. The results are a variety of optimized structures, ranging from a simple 4 x 4 meter canopy to a long-span roof (double and single plane).

Structural System

Space frames are structural systems employed when large spans are required either in one dimension (bridges, buildings) or two dimensions (roofs). As three-dimensional structures composed of one-dimensional elements, they are light, elegant, and transparent. This system was fully implemented after the industrial revolution, when steel was established as a construction material. As with any structural system developed through human history, it took some time to completely understand its idiosyncrasies. Currently, concrete (reinforced and post-tensioned) and steel are the most widely used materials for space frames.

Form Finding

In contrast to the traditional method of designing a structure, in which elements are sized given the geometry and loads, form finding has a different approach. Given the structure's loads and constraints such as stresses, deflections, and geometry, form-finding methods search for the best geometry that achieves these conditions.

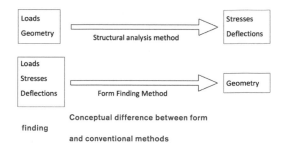

Conceptual difference between form finding and conventional methods

Methods

Numerous form-finding methods have been developed in recent decades, closely related to the development of computer-based structural software; here I will focus on two of them: the force density method and dynamic relaxation. Both are based on the finite element method.

Force Density Method

First, the space frame has to be relaxed to minimize energy while respecting several constraints such as fixed points and stresses. This method is one of the simplest algorithms developed for form finding. The principle that drives this method to convergence is the nodal force equilibrium. Every node is affected by its neighbor nodes, which subject this node to forces. In the appropriate geometry, the result of

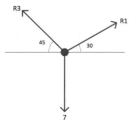

FMD node equilibrium

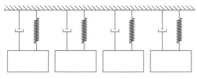

Conceptualization using spring-dashpot elements

Form-Finding: Sidney Myer Music Bowl

Soap film adopting pure tensile shape

Form-making: Henderson Wave Bridge

Gaudí polyfunicular model

these forces is equal to zero. Before convergence, however, the result will have a certain direction; this vector indicates the direction and distance to move this node.[1]

Because of the characteristics inherent to this method, perfect convergence is never achieved. Thus accuracy for convergence has to be relaxed to get final results.

Dynamic Relaxation

For this method, mass is lumped at every node, and relationships between neighboring nodes are defined in terms of stiffness. As a result, the system oscillates under influence loads until convergence. The conceptualization of the method consists of several masses linked by springs. Furthermore, influence loads are applied considering dynamic effect in the conceptual system. To accelerate convergence, adding some damping to the system is highly recommended.[2]

Difference between Structural Form Finding and Form Making

It is important to point out the difference between structural form finding and form making. Form finding uses available computational tools to obtain efficient structural shapes (pure axial forces). Many methods are available, but all base their process and therefore the final shape on the use of pure axial forces in all of the members.

Form making is the use of computational tools to add constraints to the design and rationalize, through a mathematical/physical artifact, the geometry developed. There is no objective function behind the whole process, however, and so the final shape cannot be claimed to be efficient in terms of structural design.

Pure Optimization Approach

The form-finding methods mentioned are based on regular grids to obtain tensile structures. However, this does not always favor weight optimization because of the lack of sizing included in those methodologies. To get the optimal shape and sizing through the range of possibilities, optimization meth-

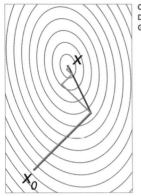

Comparison between Gradient Descent [green] and Conjugate Gradient Method [red]

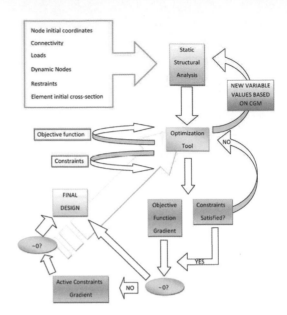

Program flow-chart

ods define an objective function (usually volume, price, or any efficiency coefficients desired for the structure), which will drive the design/optimization process. Among the optimization techniques are optimization by differential calculus (leads to minimize the weight or the cost of an element by solving differential equations), optimization by mathematical programming using graphical methods, and optimization by search methods, such as interpolation methods, conjugate direction and gradient methods, and direct search methods.

Conjugate gradient method (CGM) is implemented in a general finite element method script to obtain optimum shapes for a range of structures. The main idea of the gradient method is to calculate the gradient of the objective function for every iteration (concentric ellipsoids) and move a certain step(s) parallel to this calculated gradient. For every step, constraints are checked. Convergence is achieved when the objective function is impossible to reduce while satisfying constraints.

The conjugate method varies step size every n iterations (user defined) and changes slightly the gradient direction to accelerate convergence. The method starts at a point given by the user X0. Following the CGM pattern, the final solution comes to be X.

Program Description
As a simplification, the typologies of structures covered were limited to reduce the number of problems solved. Next, assumptions and decisions are summarized for reader ease.

One-Dimensional Elements Considered
There is a wide available range of commercially manufactured structural members. Cross-sections can be square, rectangular, or circular. This case deals with circular, hollow sections. Available radii and thicknesses are considered during the optimization process. Because of the nature of the software (MATLAB), a limited number of variables are possible to optimize. Hence all of the elements have the same cross-section. Using a more powerful optimization tool, complete size and shape optimization could be accomplished. This is considered to be more appropriate, but better software was not available at the time of writing the script.

Loads Considered
Loading combinations increase enormously the number of constraints to evaluate for every iteration. As a consequence, a single load case of vertical loading is considered in the optimization, as it is the most appropriate for the structures considered. For the larger structures, a distributed load equal to 400 kilos per square meter has been considered. However, 3,000 kilos per square meter has been considered for the smaller structures.

Connections

Pinned connections (truss) and moment connections (frames) are considered. Some of the problems are solved as a truss, while others are solved as a frame. As a result, comparative analysis is important to understand the different behavior of these two options.

Constraints

Constraints were chosen to be deflection and stress. A 50-millimeter maximum vertical deflection is imposed, while stresses are limited to the Euler buckling load for compression (π^2 *EI/$(\beta*L)^2$) and to the steel yield stress for tension (235 MPa). β is the effective length coefficient and varies from $\beta = 1$ (both ends pinned) to $\beta = 0.7$ (both ends fixed, moment connection).

Objective Function

For every problem, structural weight is the objective function to minimize.

Results

4 x 4 meter canopy with moment connection and single diagonal

Total mass	Cross-section thickness	Radius
458.5 kg	5 mm	3 cm

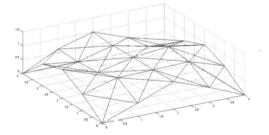

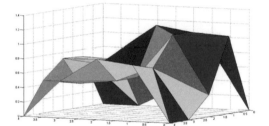

4 x 4 meter canopy with pinned connection and single diagonal

Total mass	Cross-section thickness	Radius
3,216.5 kg	5 mm	19.6 cm

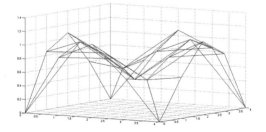

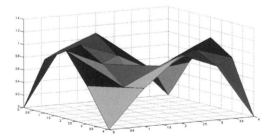

4 x 4 meters canopy with pinned connection and double diagonal

Total mass	Cross-section thickness	Radius
3,832.5 kg	5 mm	17.9 cm

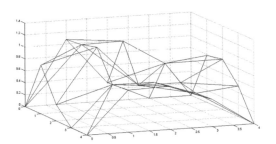 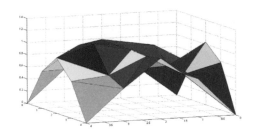

1 Layer roof with moment connection

Total mass	Cross-section thickness	Radius
10,534.8 kg	6.83 mm	3.43 cm

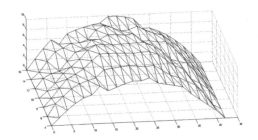 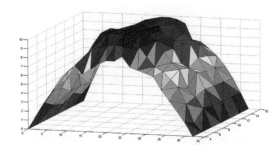

2 Layer roof with moment connection

Total mass	Cross-section thickness	Radius
15,120.9 kg	5 mm	3.28 cm

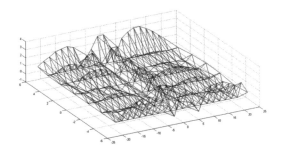 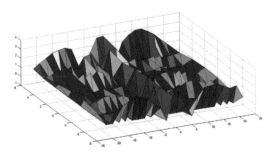

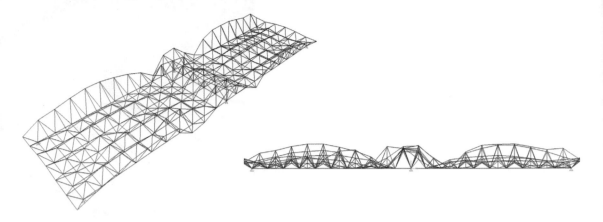

2 Layer roof with moment connection (optimized geometry). Cont

Conclusions

Comparing results between pinned and moment connections, it is important to point out that the difference in cross-section required for both is due to the difference in Euler buckling ultimate load. While a moment connection has a $\beta = 0.7$, pinned connections have $\beta = 1$. As a result, buckling load is more than two times smaller for a pinned connection. Furthermore, the addition of bending action in the members causes ultimate axial load to drop to about 50 percent. For roof structures, where members are working both in tension and compression, those characteristics lead the optimization.

When working with single-plane structures, pinned connections proved to be much heavier (with almost seven times more material). As a result, moment connections are the best choice for single plane, while pinned connections could be efficient in structures with more than one layer.

Among possible improvements, sizing element by element seems to be the most sensible next step.

In space frames, diagonals require larger cross-sections in the area close to the supports than in the middle of the span. However, concerning horizontal struts, the middle span elements have the largest cross-sections. Another improvement would involve including smoothing functions to get better results in term of transitions between neighbor nodes. One example would be to constrain the difference in height between neighbor nodes. Another way would be to adapt the structure shape to a three-dimensional NURB (non-uniform, rational basis spline) defined by certain control points. The latter would be highly efficient, reducing the number of variables to consider because only control points would be integrated in the optimization.

The optimizations accomplished did not include geometrical constraints. However, every project has specific geometrical constraints that need to be implemented in the optimization (minimum clearance, architect desire, maximum height, differential height between neighbor nodes, etc.). This aspect could be additionally implemented for any problem given.

Notes
1. H. J. Schek. "The Force Density Method for Form Finding and Computation of General Networks," *Computer Methods in Applied Mechanics and Engineering* (vol. 3, 1974), pp. 115–134.
2. W. J. Lewis. *Tension Structures: Form and Behaviour* (London: Telford, 2003).

Henderson Waves Bridge

IJP / AKT

Henderson Waves: A Collaboration

George L. Legendre

In 2004, the newly founded London-based IJP Corporation began exploring in a systematic manner the natural intersections of geometry, space and computation. Our first two commissions turned out to be large infrastructural interventions with complex structures and simple briefs. We started off fast, but there was a catch: with its simple structures and complex briefs, architecture, according to Robert Venturi in the opening pages of *Complexity and Contradiction in Architecture*, works exactly the other way around. To stay on the safe side, we gave ourselves a couple of years to understand our instruments before reversing priorities (the question is still open at the time of writing).

The brief of our first commission, Henderson Waves, was simple enough: to link two ridge summits with a plane. Our site was part of a 9-kilometer chain of hills stretching from Mount Faber Park through Telok Blangah hill and Kent Ridge park, on the southern coast of the Island of Singapore. The master plan called for proposals to link the Southern Ridges, for people to walk or cycle through the chain of hills and explore the parks as well as enjoy panoramic views. The international bridge design competition launched in 2003 proposed two alternative sites of intervention and called for two proposals of drastically different size. Picking one was easy enough: given the extreme relief and dramatic qualities of the setting, we opted for the "big leap" of Henderson Crossing over the "long crawl" toward Alexandra Link.

In urban design terms, the proposal was simple too: it was to spring from a scenic point off Mount Faber in the foreground and land on the Southwest side of Telok Blangah hill, spanning some 700 feet over a freeway. Our chosen location for the springing point maximized the visual impact of the structure over the gorge while minimizing its length. The proposed structure would bend gently toward the east and wrap around the peak. It would then reconnect with the ring road, creating a continuous circulation loop before heading west toward Alexandra Link. At the base of the loop, pedestrians walking across the bridge and moving west would pass under the deck to discover its underside. On the first site visit in July 2004 with AKT Director Paul Scott, we found that everything was buried in equatorial undergrowth, and that it was beautiful.

Bridges have simple briefs, but our internal brief was even simpler than that: to leap over the gap—and generally speaking, to do everything—with only one equation. This was actually possible, provided that one caved in a bit and subdivided the domain of application of our functions into discrete portions, to address issues as they came. The equation seed we eventually went for was a direct application of IJP's research in periodic transformations, as summarized in *IJP: The Book of Surfaces*—where, true to form, the surface was described only verbally and not rendered into a shape. We chose to work with the parametric pillow, a 3D surface obtained by composing one linear force with two periodic ones.

Mesh density and form variables PLEATED CANOPY 0.2

$i := 0, 1 .. \ 100$

$j := 0, 1 .. \ 1$

$\Pi_{i,j} := \dfrac{i}{100} \cdot 80$

$\Theta_{i,j} := \dfrac{j}{1} \cdot 16$

$K_{i,j} := \left[\cos\left[\left(\dfrac{i}{100} \right) \cdot 1.5 \cdot 2\pi + j \cdot 0.5 \cdot 2\pi \right] \cdot 10 + \cos\left[j \cdot 2\pi \cdot \left(\dfrac{1}{2} \right) \right] \cdot \sin(j \cdot 2\pi \cdot 1) + \cos\left[\left(\dfrac{i}{100} \right) \cdot 2\pi \cdot \left(\dfrac{100}{2} \right) \right] \cdot \left[-\sin\left[\left(\dfrac{i}{100} \right) \cdot 2\pi \cdot 1.5 + \dfrac{3\pi}{2} \right] \cdot 4.3 + 6.1 \right] \right] \cdot 0.22$

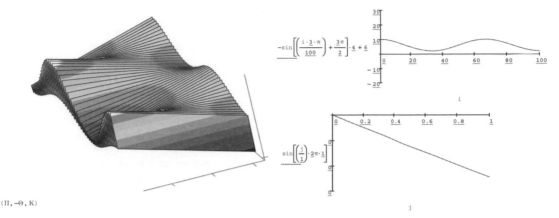

$-\sin\left[\left(\dfrac{i \cdot 3 \cdot \pi}{100} \right) + \dfrac{3\pi}{2} \right] \cdot 4 + 6$

$\sin\left[\left(\dfrac{j}{1} \right) \cdot 2\pi \cdot 1 \right]$

$(\Pi, -\Theta, K)$

Mesh density and form variables PLEATED CANOPY 0.2

The pillow is a classic product of three forces: the first force may be diagrammed as an oblique plane. The second is more complex and gives us a series of periodic oscillations whose surface plot flows in the wavelike course of a sinusoid. And the third (and hairiest one), is a product of periodic out of phase oscillations that spread in perpendicular directions. Its diagram looks like a weave-like arrangement of peaks and depressions: as the ranges vary, the pillow splits like a cell into two, three, or more swelling bulges, as if tightly held in several places by knots.

Using equations, you can write anything you want. As we quickly found out, some of the results we obtained not only looked groovy, but reminded us—rather uncannily, in fact—of forms one could arrive at by more traditional means, say by using Rhino or Maya. Where there is nothing, everything is possible, but in this case, specific structural considerations drove the process. We stayed away from fancy expressions. Following Paul Scott's advice, we went for a comparatively conservative scheme that maintained the structural integrity and load-bearing capability of the surface's longitudinal lines, a form we could diagram as a succession of arches and catenaries. In hindsight, this was not a bad move, as, having won the competition, we were also required to build it.

The overall geometry of the steel structure was—and remains—far from simple: to cross the gap between springboard and landing without going over the prescribed slope, the bridge had to bend in plan, while rising steadily. It also had to undulate, and the deck had to be broken at regular intervals to allow wheelchairs to pause. As a result, somewhere in the background of the form, the periodic motions induced by the equations made it glide out of alignment in all three dimensions. It bent, undulated, and climbed by 60 feet, all in one go. In the process it also deformed and self-intersected to provide adequate egress, shelter, and scenic viewing for pedestrians, marathon runners, and cyclists. The scheme submitted in 2004 had nine waves, and the largest span, 57 meters long, flowed over the high-speed traffic of Henderson Road. At 6 meters high at its apex, one could have fit a small shop under this arch. The developed length of the deck ran over 300 meters. The difference in elevation between springing and landing points is about the full height of a seven-story housing block. This proposal was, by any standard, a large one.

The timber deck presented us with the greatest geometric challenges. Looking closely at its surface-form today, you can't help noticing the kinks at regular intervals. Some observers put it down to a mistake, but that is not the case. Like the steel structure around it, the ramp had to bend and slope, but it also had horizontal thresholds that broke its continuity. Every flight and landing of this discontinuous surface had to be computed by a specific variant of the equation, which means that every landing of the bridge, however simple its outline, had its own mathematics. The final numbers were gleaned up to the fourth decimal from fifty-five different math sheets—all held together in one master spreadsheet, and giving us the precise form of the bridge's "flight path" across the valley. Admittedly, the jury is still out on the issue of efficiency, but when we did this in 2005, we weren't practicing architects just yet; while doing this, we felt more like air-traffic controllers. We did not know what the tolerances of the building industry were, and we produced the first disability-access ramp ever designed exclusively with analytic geometry.

Is such an emphasis on geometry misplaced? While visiting the exhibition organized by the Government of Singapore to publicize the competition shortly after announcing the winners, I noted among the schemes submitted by competitors a rather large amount of simple, straight bridges. That was surprising. No linear path sloping between the designated areas on either side of Henderson Road would have been shallow enough for people to walk on, let alone haul themselves up on a wheelchair: it was simply impossible. Simplicity was a deceit. Given the accessibility and topographic requirements handed to us in the brief, the double curvature was not a matter of fancy, but of ruthless practical necessity.

Structure and Supports

In terms of structural expression, Henderson Waves demonstrates the principles of the abstract physical Indexical that Models IJP has developed over the years. It systematizes and amplifies the engineering problem of converting selected indexical threads into structural members—with specific structural roles. Structure and surface morphology are closely linked.

The transversal threads (or iThread) serve the physical construction of the surface-form and ensure the lateral stability of the bridge. These profiles are flat and parallel to the others, which simplifies manufacturing. The rotational and mirror symmetries of the pillow provided us with another key benefit: the texture of the surface-form is largely redundant. Of a total of 498 curved profiles, only 81 have unique dimensional features—with the remainder being affordable copies. The protocol of material translation was straightforward enough: threads 73 and 81 were structural steel box sections (83 in total), and all threads lying in between (415 in total) were secondarily structural i-beams.

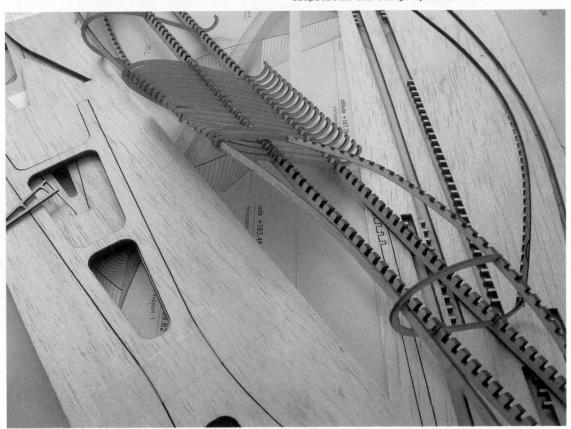

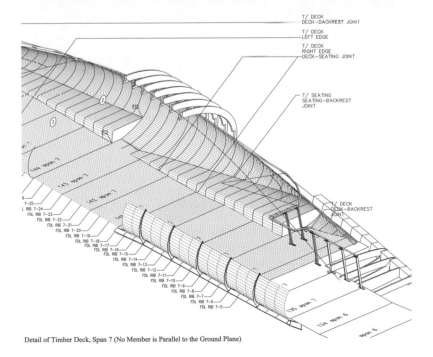

Detail of Timber Deck, Span 7 (No Member is Parallel to the Ground Plane)

Along the other range, the longitudinal threads j fulfil the essential demands of the primary structure: Threads 1 and 20 provide a central spine member, giving support to the deck at midpoint, and the edge member that supports the open area of the deck. Thread 45 supplies the central arch member that rises and falls in alternate spans, in turns arch and catenary. Thread 26 became the mid-height member. This key piece forms a shallow arch that provides support to the curved steel profiles. It alleviates the need to restrain the arch above with a larger spine member working in bending, and accounts for the nearly implausible thinness of the overall structure.

These members are all doubly curved and required precise development. To model them precisely at scheme design time, we built a primitive development machine (we thought it would help us bend basswood members in three dimensions, but it didn't work, although we did manage to build a scaled model of one wave, or about one tenth of the bridge; even at this scale the full model would have spanned over 3 meters).

The relationship of Henderson Waves to the surrounding landscape was always blunt, and from the beginning, this put a lot of pressure on the piles. We decided against a literal continuity between superstructure and supports, mostly because of the latter's scale: the tallest pile rising up to 40 meters, which would have required structural gymnastics we had no inclination for. We concentrated instead on including them carefully within the overall periodic pattern. From the beginning, we positioned the supports at the point where the surface self-intersected, along a line of inflection where the upside wave turned into a downside one, where the surface dwindled to a single curve, and the mass of bridge came down to a single beam. Not being generated by the equation, these concrete supports had to be actually designed and quickly established themselves as a most controversial feature. We produced seven designs in total, all inconclusive. After a sophisticated option based on branching nodes that had been approved but had to be thrown out, late in design development, on account of potentially dangerous harmonic vibration patterns, we simply gave up and picked the simplest option. The client had grown fond of the branching option and had to be convinced to accept the simplest alternative. The project stalled for a few months, and may have even been under threat. The saga of the bridge's vertical supports offered us the only—yet starkest—reminder that we were not designing a building.

Deck

At the competition stage we had envisioned the super-surface as a kind of thin timber veneer stretched over the steel members. Within the side pockets, the super-surface was to seemingly peel of the structure to provide seating and playing areas. The transition from form to program would be direct: Pedestrians and cyclists wishing to pause to look at the landscape, seek shelter from the rain, or witness the passage of runners would sit on the folded areas. Eventually the form of the veneer was given to us by the same overall equation, with minor adjustments for the seating areas, which required their own custom calculations.

The 1,500-square-meter timber expanse became the centerpiece of the project. To help establish the sustainable nature and origin of the tropical hardwood yellow balau, we consulted with the Singapore-based timber supplier and global environmental activist Certisource from the earliest stages. The collaboration was most successful. The complex, doubly curved portions of this large expanse of tropical hardwood form a tapestry of 5,000 modular boards, each varying by a single degree every 10 meters—and all tapered to measure. In addition to sourcing the timber through Certisource, Venturer managing director Kevin Hill oversaw the fabrication and installation of the timber components and conceived the timber details and specifications through a combination of modular fabrication and painstaking installation work. The manufacturer's custom installation techniques (dowel-fixed with stainless pins and epoxy filling) guaranteed the absence of visible fixings on the entire sculpted timber surface (especially at the rear backrest, which is doubly curved).

With the entire deck supported on a steel subframe with vibration dampeners, the coordination of steel and timber became a challenging task. Using our proprietary equations, we issued precise numerical descriptions of the surface at regular 500 millimeter intervals and provided dimensional coordinates that assisted the production of the timber manufacturer's shop drawings. To correct any variations between the timber surface and the outer face of the steel caused by the galvanizing process, deck panels were joined with mastic. The timber work was carried out by timber craftsmen from mainland China, under the direction of Venturer's director and site manager, Max Pereira.

Collaboration

Like all realized projects, Henderson Waves is the product of a complicated collaboration. Henderson Waves gave IJP its first opportunity to nurture its instrumental priorities in the form of an open dialogue with its engineering colleagues from Adams Kara Taylor. The dialogue has nonetheless been open but tricky, as the engineer can be a closet archi-tect—and the architect a closet engineer. Each has something the other party clearly wants, but neither is willing to give up the perks and priorities of their discipline in the process.

There are other demarcation lines. Architects who do not understand the physical mani-festations of structure (physical understood here in the sense of physics) may have a good sense of its mathematical nature, an equally complicated process that does not always translate into material or sensible things. This is why our colleagues at Adams Kara Taylor are occasionally baffled by our inability to intuit the arithmetic of bending moments, when we are so comfortable with the integration and differentiation of complex equations. This apparent contradiction has nothing to do with adequate preparation or skill; it simply comes down to attitude. At heart, the mathematical nature of things is only a game (albeit a complex and risky one, at which one's academic standing, self-esteem, and career options may be gambled away ridiculously early), whereas the physical nature of things is simply a necessity, that is, an inevitable fact of (material) life: without it, buildings would collapse, and we would have nowhere to live. Architects seek out the artificial nature of complexity rather than its physical nature, and in this sense, with its continued emphasis on factual, material, structural, and economic behaviors, the great maturity of the engineering profession may forever elude us. Our contrasted priorities also help explain other apparent contradictions, such as the casual attitude our engineering friends display toward our shared disciplinary inheritance, mathematical geometry. To the enlightened engineer, mathematics after gradu-ation is merely a tool, banished to the fringes of the discovery process and increasingly con-fined to vocational training or calculation software. To the enlightened architect on the other hand, mathematics is not a tool, but an idiom, a language to speak in—and the engineer's conceptual speculations hardly a match for the powerful vehicle of discovery this language affords us, untrammelled by the need to be immediately evaluated and understood. Like all key choices in life, what we might call the "architectural and engineering attitudes" have contrastive benefits and trade-offs.

Coda

In the early years, IJP did not release any renderings of the bridge. Being iconoclasts in a literal sense, we avoided all imagery, especially of the synthetic kind (and even more so of synthetic imagery meant to depict the mathematical surface). What served us well in research (where our studios and design experiments were always rendering-free), turned out to be problematic in practice. Throughout the first four years, Henderson Waves received little coverage in Asia, even though the images of physical models we routinely sent to journalists reproduced all the tropes of computer imagery: the reflection, the dynamic camera angles, the dramatic lighting. Our correspondents just couldn't relate to the conventional nature of these images any more than they could understand the hundreds of drawings produced. I suspect that they were finding us a bit unprofessional, perhaps even a bit incompetent. There was some kind of misunderstanding going on, but back then it didn't matter. We had waited long enough to get to this point, and could wait some more. In late 2006, the project entered a brief phase of difficult negotiations, and things got a bit tangled for a while. Construction began within eighteen months, and the project was successfully delivered in 2008. Henderson Waves received the Singapore President's Design Award 2010. The project has welcomed 200,000 visitors to date and is featured on more than a million websites.

Notes

1. Robert Venturi. *Complexity and Contradiction in Architecture* (New York: Museum of Modern Art Press, 1966).

2. George Liaropoulos-Legendre. *IJP: The Book of Surfaces* (London: AA Publications, 2003), p. 28.

3. http://www.certisource.co.uk/ Assistant Professor Chew Fook Tim, Department of Biological Sciences, National University of Singapore. Also of interest an online article published around the time of IJP's consultancy with Certisource, which uses DNA test to identify legal logs: http://www.illegal-logging.info/item_single.php?it_id=2028&it=news 15/04/2007.

Double Shell

YueYue Wang, Hailong Wu,
and Zhu Wu

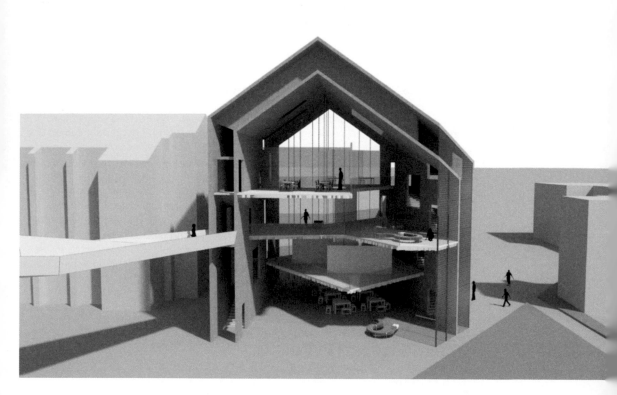

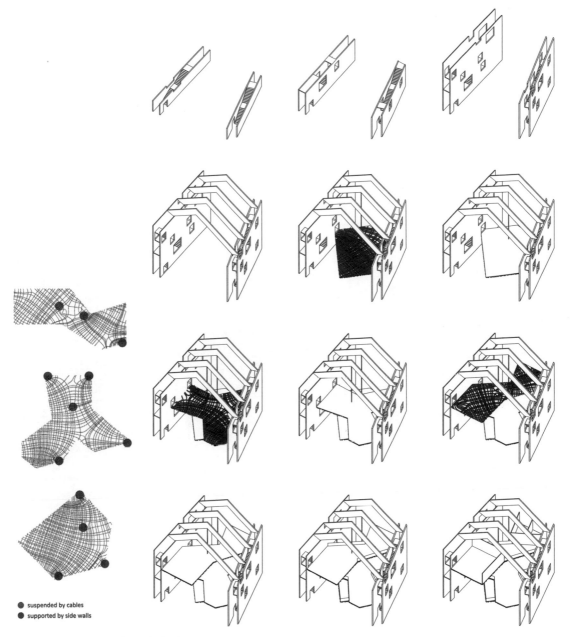

- suspended by cables
- supported by side walls

Overall Concept

How can engineering benefit design? Instead of responding to the question from the perspective of what engineering can do, we are more interested in what engineering should do. Of course, engineering could express itself in a technical way, but what if we minimize the expressions of technology and emphasize the environment and spatial atmosphere that the engineering creates?

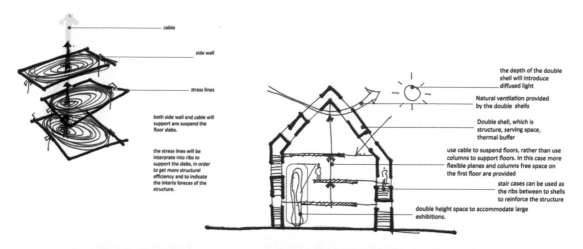

Structure

The structure of the building is to provide a highly flexible space with variant environmental configurations. The initial idea is an entire high space with slight horizontal subdivisions. This could accommodate the unpredictable program, such as gatherings and exhibitions of different sizes. To achieve this goal, we introduce a structure system that is a megastructure conceptually, but a light one experientially. It starts from a double-shell frame that hangs over free-form boundary floors with two cables. Then we need to make the structure as invisible as possible to make users focus on the place it creates, rather than itself. We make openings in the shells and add ribs between them along the boundaries of the openings. In this way the structure becomes transparent visually, but reinforced structurally. By locating supportive functions within the space created by the double shell, we maintain a clearer span for the main program.

Context

The specific design process starts from opportunities presented by the site, located in the backyard of GSD's Gund Hall. First, two shortcuts cross the yard, with many GSD students passing by every day. Therefore we make the south façade next to the shortcuts open, with the entrance atrium beside it. Second, we set back our building footprint to maintain the completeness of the yard, which could be an extension of interior public activities. The transparent glass curtain wall to the south maximizes the visual connection between the yard and the interior space.

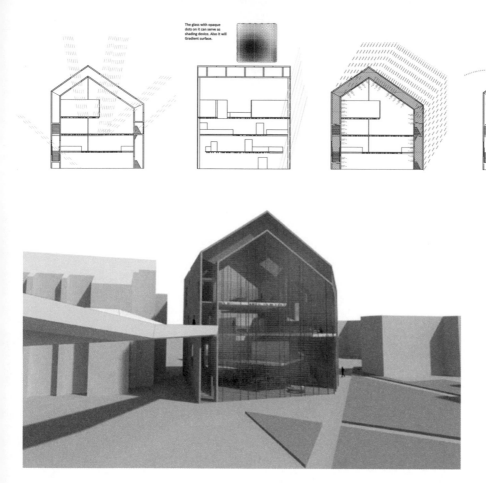

The glass with opaque dots on it can serve as shading device. Also it will Gradient surface.

Ecology

The glass curtain walls on the north and south sides bring enough light to the public atriums. On the south side, we set back the first and second floors to form an atrium, covered by the third floor, which can serve as shading for the atrium in summer without blocking sunlight in winter. On the east and west sides, openings in different locations and of several sizes on the shell introduce natural light and produce a varied lighting environment.

– In winter, the cavity between the double concrete walls can serve as thermal insulation. Furthermore, it can help in collecting heat from sun. Because of the high heat capacity of the concrete, the outer shell, the cavity, and the inner shell will be heated successively by the sun. Heat radiation from the cavity and concrete could help warm the keep interior space. In summer, the concrete roof can serve as thermal mass to flatten out natural daily temperature fluctuations.

– In summer, we can open the windows on the outer shell to invite the natural wind to flow through the cavity in the roof area. The resultant low wind pressure of the cavity will drive the natural ventilation of the interior space, pulling interior hot air out.

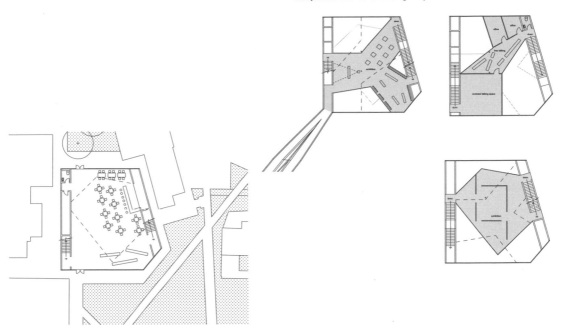

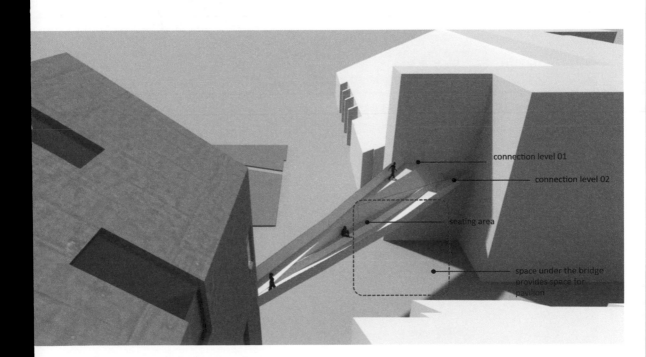

connection level 01

connection level 02

seating area

space under the bridge
provides space for
pavilion

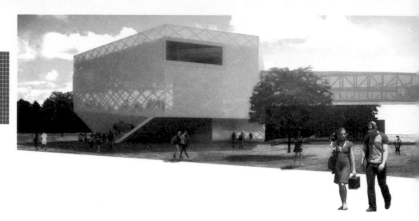

Jewel Box

Vera Baranova, Sophia Chang,
and Bernard Peng

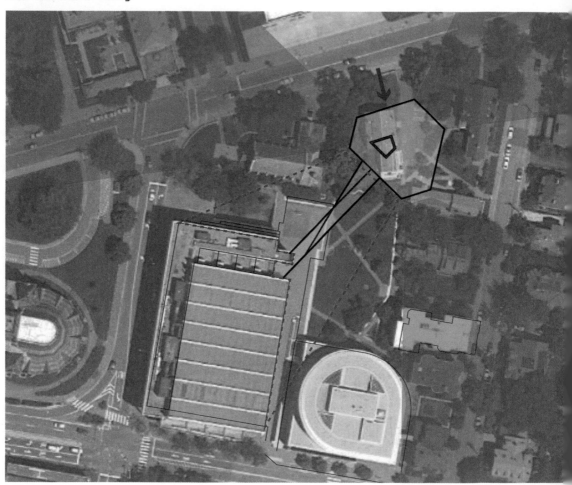

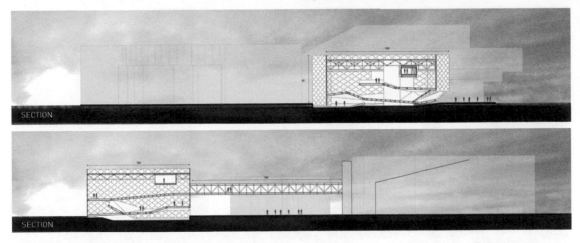

SECTION

SECTION

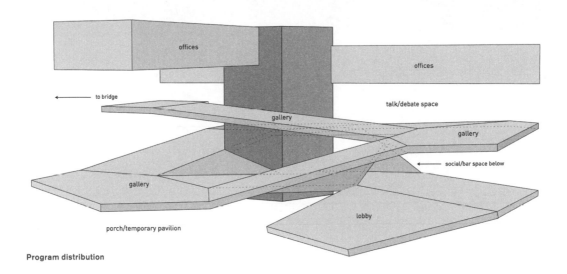

offices

offices

to bridge

talk/debate space

gallery

gallery

social/bar space below

gallery

lobby

porch/temporary pavilion

Program distribution

The project sits in the backyard of Gund Hall and frames four views. The main view focuses on Gund Hall; it occurs at the bridge connection between the buildings. Since the connection from the GSD occurs through the bridge, a public entry exists on the north side, on Kirkland Street. A ramping path wraps around the core, encompassing a continuous exhibition space and connecting the public entry to the bridge above. A view is framed by each of the ramp's three platforms, looking out toward campus and back toward the neighborhood of Somerville.

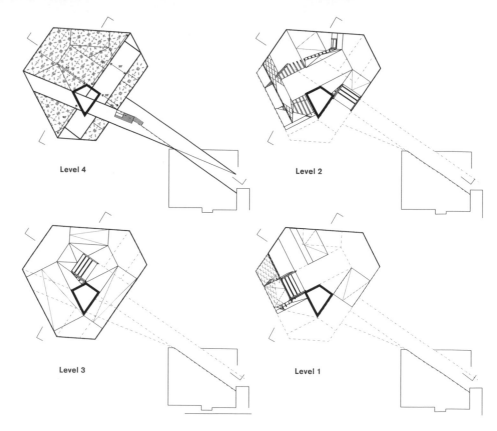

Level 4

Level 2

Level 3

Level 1

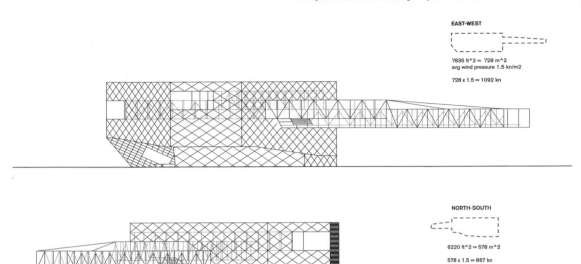

EAST-WEST

7835 ft^2 = 728 m^2
avg wind pressure 1.5 kn/m2

728 x 1.5 = 1092 kn

NORTH-SOUTH

6220 ft^2 = 578 m^2

578 x 1.5 = 867 kn

Program distribution

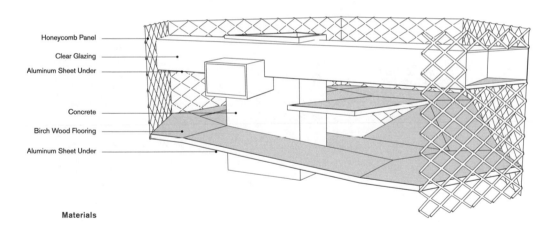

Honeycomb Panel
Clear Glazing
Aluminum Sheet Under

Concrete
Birch Wood Flooring
Aluminum Sheet Under

Materials

These views are framed additionally by the ends of the office cores, which act both as cross-bracing and as part of the three-dimensional truss at the roof. This truss sits on top of the core, pinned down by a structural diagrid to the north to support a large cantilever over the porch space facing the GSD backyard.

SOUTHERN FACADE

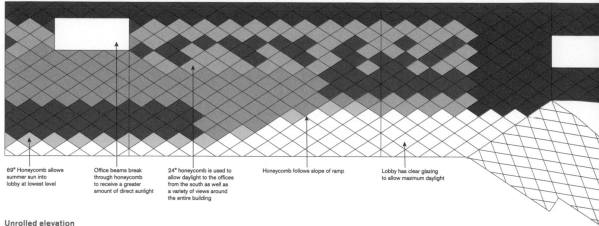

69° Honeycomb allows
summer sun into
lobby at lowest level

Office beams break
through honeycomb
to receive a greater
amount of direct sunlight

24° honeycomb is used to
allow daylight to the offices
from the south as well as
a variety of views around
the entire building

Honeycomb follows slope of ramp

Lobby has clear glazing
to allow maximum daylight

Unrolled elevation

The diagrid is filled with a designed system of insulated honeycomb panels. To push forward this new product, a prototype will be made, patented by the architects, and sold to a manufacturer. During manufacture, the honeycomb will be cut at a various angles to create different opacities, controlling daylight and views into and out of the space.

NORTHERN FACADE

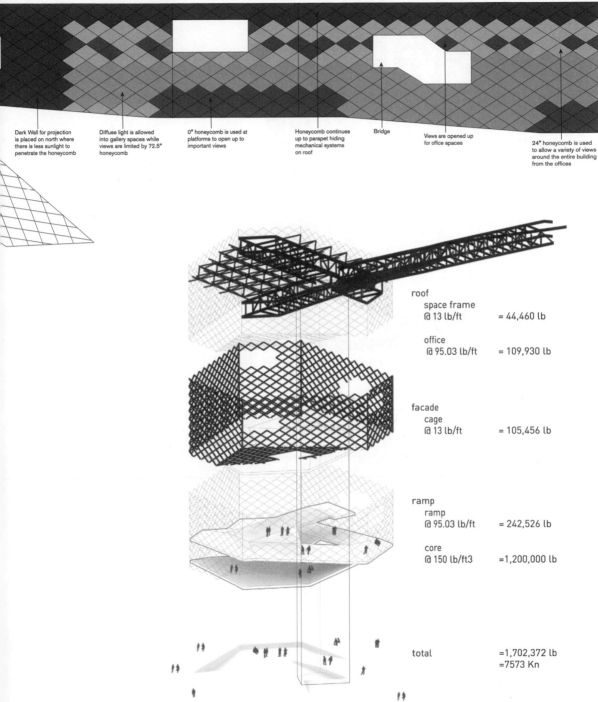

Dark Wall for projection
is placed on north where
there is less sunlight to
penetrate the honeycomb

Diffuse light is allowed
into gallery spaces while
views are limited by 72.5°
honeycomb

0° honeycomb is used at
platforms to open up to
important views

Honeycomb continues
up to parapet hiding
mechanical systems
on roof

Bridge

Views are opened up
for office spaces

24° honeycomb is used
to allow a variety of views
around the entire building
from the offices

roof
  space frame
  @ 13 lb/ft   = 44,460 lb

  office
  @ 95.03 lb/ft   = 109,930 lb

facade
  cage
  @ 13 lb/ft   = 105,456 lb

ramp
  ramp
  @ 95.03 lb/ft   = 242,526 lb

  core
  @ 150 lb/ft3   =1,200,000 lb

total   =1,702,372 lb
                            =7573 Kn

Structural axonometric + weight

UK Pavilion
Shanghai Expo 2010

Heatherwick Studio / AKT

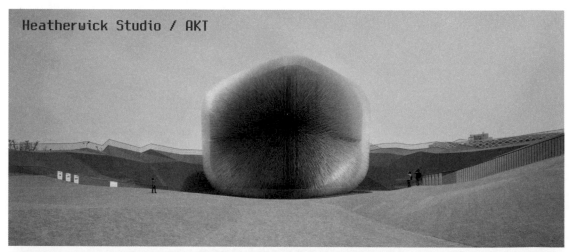

In Collaboration...

Katerina Dionysopoulou

At the early stages of our architectural lives, when still in academia, we learn how to approach a "project." Regardless of the scale, location, purpose, usage, and complexity of the "project," we learn how we as "architects" should carry it through:

> We learn the importance of the "idea," and how to respond or form a brief.
> We learn how to come up with a concept and translate it into a design.
> We revisit the brief, ask questions, analyze and filter the information, then adjust and readjust our design accordingly.
> We use our theoretical background and research to examine the urban, social, political, and economic parameters that could affect the outcome of our work.
> We draw, redraw, think, and visualize our design in both two and three dimensions.
> We learn how to incorporate mechanics and technology, and how to tackle environmental issues.
> We suggest innovative, sustainable, and aesthetically pleasing structures as part of our design.
> We outline the look of the exterior, the interior, the lighting, the content, the nature, and the very life of our "project"—the way it's used, the experience of the visitor/dweller (what he sees, how he moves), and we even try to predict how he will react and interact with our design.

What we don't learn is that the involvement of people whose professions and expertise is inextricably linked to the architectural work is essential for the successful completion of a project. We don't learn about the interdisciplinary communication that is necessary for every aspect of a scheme and the tools and methods that can be used to enhance our work with the valuable input of consultants.

Student work is the outcome of a process solely affected by the architect, and thus we never discover the challenge that lies within the complexity and intricacy of initiating and leading a conversation among a diverse group of professionals. We think and design in an environment that is purely architectural; we argue our ideas with fellow architect-students and try to communicate them to our studio leaders, who act more as clients and focus on guiding us through a rather independent and less inclusive design process. That of course helps us become better designers, builds our critical thinking, sharpens our senses, and gives us a wider perspective on the architectural design front. Our tutors prepare us for the client-architect reality; they give us a brief and challenge our response to it; they evaluate our product, question it, and either celebrate or reject it. The necessity and value of the consultation mechanism is not fully conveyed.

When an architect starts working on real projects of a certain scale and complexity, it becomes increasingly evident that his or her role is not as centralized and the process not as linear. Professionals from other disciplines are essential in all three phases of concept, design, and construction of a project. The sometimes isolated environment that architects may enjoy working in is penetrated by the input and expertise of those professionals. This dialogue between the different disciplines is hard to structure, sustain, and most important, lead and is always a challenge for the architect. When carried through successfully, though, the result is optimized. At this point, the people involved have maximized their input and learned from the creative interaction.

In some cases, the same university approach described above is carried through to the professional environment, with offices allowing or even forcing architects to design in isolation. Following that working model, architects design and validate their work internally. Externally they focus only on cultivating the desirable relationship with the client, one that will allow them to carry their design through from concept to construction with no surprises. They bring in consultants at late stages and deal with them as if external forces that try to alter their precious work and thus do not integrate them in the process. The design, the structure, the services, and all of the components that make a project come to life are not woven into a consolidated piece of work, making the project approach fragmented and weak.

In other cases, the clients feel most strongly about the architectural practice that they have chosen to work with and neglect the consultants. That creates a similar unbalanced situation where the team doesn't work, think, and act as one. A hierarchy is imposed on every aspect of the communication and the way the work is organized and carried out, resulting in limited performance by every party.

In some offices, however, consultants are invited to participate from the initial concept phase. This helps as ideas are formed collectively, through dialogue. When conversations and concept design sessions occur, ideas from people with varying backgrounds are heard, noted, implemented, deconstructed, and put back together. Such a process results in more solid and multilayered concept. Throughout this process, from concept phase to completion, people from different disciplines are given the chance to think outside of the strict boundaries of their professions, exploring and reestablishing them. With this method, professionals are challenged to understand and embrace different points of view, improve their way of thinking, and widen their perspective and approach to design. At the same time, they learn to appreciate, listen, and take a positive stance toward people from other disciplines and their contributions to the project. The mutual respect that team members exhibit creates an atmosphere where the energy and speed of the creative process benefits everyone. The outcome is a consolidated piece of design where every element of the concept is thought through.

Apart from the interpersonal and creative benefits of an architect-consultant collaboration, there is also major optimization of efficiency. Less time is wasted in ineffective design sessions where everyone tries to force their ideas and ignore or reject others'. Everyone's involvement and understanding of how everything works and why everything is designed as it is, from an early stage, also reduces challenges during construction.

Competition image

That, in combination with an overall team understanding of the budget and the constraints of every stage, produces a smoother process and a more successful result in terms of time and money. A strong and challenging design sometimes acts as a bond that brings the team together, motivating architects and consultants to push the boundaries of their fields and create something innovative and extraordinary.

The British Pavilion

In designing and building the British Pavilion for the Shanghai Expo 2010, many of the issues and processes mentioned above became a reality. Engineers, environmental analysts, lighting designers, and interior designers worked closely with the architect to decide which materials got used and to address the details, the incorporation of the structure in the design, and the life and potential behavior of the building within the context of an expo and within the larger geographical area of China. Material, climate, structure, size, and usage issues were adjusted and readjusted. Design explorations and research into possible solutions were happening simultaneously. The contractor was invited to be involved from the early stages of the project. This brought a pragmatic perspective to innovation from the beginning of the team's thinking process. That minimized the surprises and challenges the team faced during construction.

The sociopolitical agenda was also addressed in the early stages of design, as project managers, exhibition designers, content advisors, traffic flow analysts, and branding designers were all included. The visitor experience was a key aspect of the project and the way it was envisioned, affecting the landscape design, the circulation around the "Seed Cathedral," and the way the interior spaces were designed.

The team began working together from the concept design phase, which started with the architect's initial idea of the soft, hairy object that would respond to environmental conditions and sway in the breeze. The building had to convey a range of ideas, and through the team's collaboration it all came together as a consolidated piece of work. Through research into how people react to impressive structures and the wealth of ideas in an expo environment, we decided to create a structure where architecture and content would be inextricably linked. The aim was to offer one extraordinary experience that would turn into one solid memory—no fragments and glimpses of separate buildings, exhibitions, ideas, and spaces. We focused our resources on the development of one amazing object.

Having entered the competition as a consortium, Heatherwick studio, AKT, Atelier 10, and Casson Mann won based on these ideas. We presented a series of diagrams that would communicate the ease of fabricating this object and erecting it on site, managing to sell to the public sector and government something that would normally sound risky—a pavilion that would sway in the breeze. We worked consistently as a team for the next year to convince all stakeholders that even though our unique structure came with a lot of challenges, it would meet everyone's expectations and offer a unique visual and emotional experience to visitors.

We worked on the initial idea for more than six months, collaborating with the engineers to find the appropriate materials that could work structurally and also be specific to the content we wanted to exhibit. Following a series of conversations about the pavilion's content, we decided that we wanted to communicate one main aspect of the country, which was its relationship with nature. The United Kingdom has always integrated nature into its cities. We found out that the first public park of modern times opened in Derby a year before the Royal Botanical Gardens of Kew opened their doors to the public. We approached Kew, as we wanted to collaborate with them and exhibit seeds from the collection of the Millennium Seed bank.

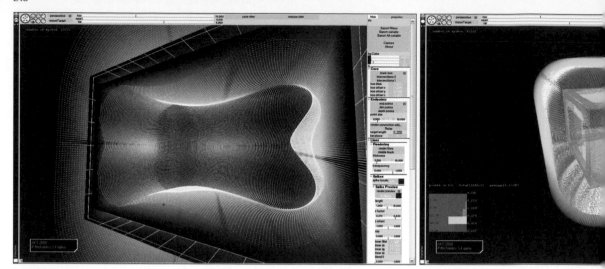

Analysis diagrams

In the process of developing the hairs as exhibition cases that would carry the content of seeds yet allow light to enter the space and illuminate it, we focused our research on transparent materials that could act like fiber optics, such as acrylic, polycarbonate, and glass. With AKT, we did series of tests to understand the properties of a number of materials that would represent the pavilion's "hairs." Tables of analysis guided us through options, which would affect the hairs' length, thickness, and proximity. The tests were initially carried out in the United Kingdom and later in China; finally a material was chosen and incorporated into the overall design.

The engineers approached a company that would run a series of tests to validate its structural abilities and its exact properties. In parallel, they developed a program that worked out the geometry of the pavilion based on a set of structural and aesthetic parameters. This program produced the drawings necessary to build the pavilion's structural components. Together with the architectural and mechanical drawings, they formed the construction set that the contractor would use to fabricate the elements.

At the end of the development phase, the soft, hairy object had became a structure that encased the content and hovered above an unfolded, textured landscape, which would act as a park for visitors. Working with seed morphologists from the Royal Botanical Gardens of Kew, seeds found a new home at the tips of the transparent hairs and gave their name to the hairy object: A Seed Cathedral.

During the design development period numerous challenges appeared, and finding solutions required long conversations among team members, stakeholders, and the client. The aim was to exhaust all options to be sure that the decision made was the most suitable and always integrated with the design. The team tried to avoid finding solutions that would make the project look fragmented and inconsistent or subvert the main goal of producing a unified piece of design.

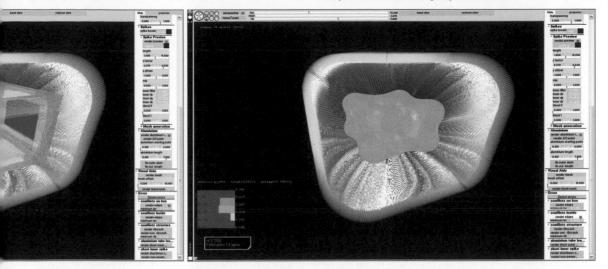

Following the client's decision to construct the pavilion in Shanghai, the team traveled to China to find contractors and fabricators. We worked with a contractor and built a piece of the structure to make sure that our intent had been communicated and also to develop some of the details in collaboration with them—making them feel part of the team through this process and embracing their approach. The material tests were repeated to reassure the Chinese authorities that the pavilion would be structurally sound and able to welcome 8 million visitors in just six months.

The team was in daily communication to make sure that all necessary tests were carried out appropriately to minimize risks. Close collaboration with the contractor was key, as it helped us to identify solutions that would ease the budget and facilitate construction while assisting with the build methodology. Two notions were key in our design concept: "buildability" and innovation. The consultants (and specifically the engineers) tried to embed these ideas in every aspect of the project at every stage. Collaboration with the local design firm was also essential in ensuring that our design followed local rules and regulations.

During the construction phase, when the main part of the team was formed in Shanghai—the contractor, project manager, and local design/construction firm representatives—the engineers and other team members remained in London. The architect traveled monthly to Shanghai, attending meetings with the local authorities, fabricators, and contractors in China. What was crucial at this phase for us, as the consortium representatives, was to coordinate, lead, and direct these conversations that were happening on a challenging geographical scale. Understanding and translating everyone's ideas, communicating the structural and environmental values of the design, and tackling the social and political issues was part of our role at this stage.

Exterior construction photo and assembly of spikes

Building a pavilion that represents a country in the largest expo to date put a lot of pressure on everyone, and there were many challenges along the way, as there are when building something that has no precedent. It required a team consisting of people who were intelligent, knowledgeable, open minded, and willing to make the effort to understand the process and see the potential. It required a client prepared to trust the architect and the consultants, and believe in creating something that unique yet functional. The result was an object where architecture, structure, and content are so inextricably linked that people couldn't understand where one finished and the other begins. Every detail traced back to the original idea, and although everything was unique, it was all cleverly designed to be cost efficient. It won the gold medal of the expo, and although the journey was challenging, at the opening day the whole team was there, united and proud of what had been achieved.

Interior construction photograph

Interdisciplinary design is a function of simultaneous action around project-level and organization-level processes, values, and investments of relentless energy. Whereas the previous sections focused on the project-level analysis of what interdisciplinary design could do, this section will focus on what interdisciplinary practice could be.

The Emergence of a New Discipline

Death of The Star Architect

Jennifer Bonner

Cassiopeia A: Death Becomes Her

"Architecture in crisis"–an overly familiar claim when assessing the profession's current state. To understand our present-day predicament, the following argument looks at one particular crisis in architectural history: Alberti's establishment of architecture as a profession and the subsequent birth of the phenomenon known as the "Star Architect." In what ways do stardom and the star system diminish the value of architecture? Is the grand narrative finally over? How might emerging productions of architecture offer transformative clues in the field? According to our agreement with the profession and its system of celebrity, let us first define the terms and conditions.

Terms and Conditions of the Star Architect
The following terms and conditions of the Star Architect are here by incorporated in and made part of the Agreement for Professional Services (meanwhile celestial analogies concurrently demonstrate the limits of stardom):

1. The Star (hereafter referred to as "Star") Defined. In the context of Hollywood's television and film industry, a Star is someone whose fame is bright and all-consuming—recognized widely throughout national and global media, in which the name is synonymous with visibility. Publicizing the actress or actor "starring" in the film is central to the marketing strategy where the Star is a primary object of interest. The "star system" involved the creation of glamorous identities and even artificial personas, in which names and personal histories were fabricated ex nihilo. Viewers are enticed to watch a film based solely on the known Star, for the image, not necessarily out of interest in the production or even the genre.[1]

2. The Star Architect (hereafter referred to as "Star Architect") Defined. A version of this celebrity status is also found within the architectural profession. The Star Architect is an individual who generates media attention through the promotion of writings, lectures, philosophical agendas, and built works. Similarly to the Hollywood version of Star, it is not the quality or production of architecture on which value is placed. Rather, it is based around a perverse but pleasurable obsession with a central figure. Collectors and clients concerned with image commission Star Architects solely based on the visibility their name provides. This subsequently attracts visitors to those buildings who are more fascinated with the exclusivity of Star Architecture than the actual structure.[2] This system of commissioning architecture further disconnects the architect from architecture.

3. The Starchitect (hereafter referred to as "Starchitect," incorporating and expanding the terms of §2, "Star Architect," above) Defined. A clever portmanteau, Starchitect labels an esoteric subculture of architecture where lingo is coded and privileged members are familiar with masonic handshakes.

The invention of this term came from within the discipline of architecture. It is a self-organized nomenclature, defining qualities of how the Starchitect envisions oneself. Identifying physical characteristics of the Starchitect include carrying around thick volumes of personal monographs, sporting black clothing, and wearing large geometric eyeglasses despite still passable vision. Herewith, the Starchitect professes grand narratives in architecture (in which, of course, he is a major and defining figure) and is charged with taste making.

4. Fact. Within a celestial system, a supernova momentarily outshines an entire galaxy with increasing brightness. As seen through the telescope, visual hierarchy separates the massive glow of the supernova from that of the simple nova or star. A similar rank and hierarchy is found in architecture's star system—supernovas and stars operate on different terms.

5. Reasoning. The Supernova Architect radiates an extreme glow with timely energetic bursts, outshining fellow Star Architects. Several conditions enhance the Supernova Architect's status including number of buildings constructed, geographical boundaries surpassed, and recognition received outside of the architecture community.

6. Heretofore, differing positions regarding the role and relevance of the Star Architect (by said Star Architects themselves) are offered through the following claims:
a. Facts: A noted Star Architect argues for full exposure—if architects take a stand against participating in the discourse of the profession and engagement with a nonprofessional public—"it diminishes the value of architecture."[3] In this argument the Star Architect is an advocate for the star system, claiming that the value of architecture or its cultural relevance hinges on an association between the public and the Star. In these terms, a vital role of the Star Architect is to play the part of a ubiquitous persona in the public eye. She or he must always perform—not professionally, but as the "Architect" strategically scripted onto public stages, expelling grand narratives.[4]

b. Previous History: A Star Architect who stood next to her professional partner (who is also her husband) at the 1991 Pritzker Prize ceremony asks, "Why do architects need to create stars?" Her own answer, "because I think architecture deals with unmeasurables."[5] It should be noted this particular Star Architect is both an advocate and a critic of the star system. In these terms, further specification of the Star Architect's relevance pivots on a desire to measure the profession within intangible means. To position "The Architect" in popular society, she or he is tasked with becoming a celebrity. Hereby, in a self-initiated effort to convey measurability, the act of collapsing celebrity and architecture further widens the gap between architect and production.

c. Reasoning: A well-known Supernova Architect and critic of the star system finds the star architect nomenclature problematic, "My hope is through the current complexity that title will exit discreetly and disappear."[6] Complexity in culture, building scale, and geometry attribute to a set of terms and conditions where the singular Star figure becomes irrelevant in the production of architecture. This often-repeated undesirability of the Star Architect questions the role of the star—yet Star Architects continue to populate throughout media, academia, and the profession. Hereafter, the Architect must be discovered through a completely different set of terms to make further advancements within the discipline.

7. Fact. A supernova implodes or explodes through gravitational collapse, thus creating a multitude of new star formations. Imploding stars and supernovas suggest definitive limits of architecture's star system.

8. The Rising Star (hereafter referred to as "Rising Star") Defined. Patiently waiting to receive Star Architect status, the Rising Star operates inside familiar and safe legacies rather than questioning the star system outright. From inside the Rising Star camp, we notice a continued acknowledgment of grand narratives currently being played out within the discipline:

a. The *Political Activism* of practices such as those responsible for making border territories hip and infrastructure not merely a civil engineering mantra.

b. The *Branded Entrepreneurialism* of figures formerly trained by Star Architects, armed with cartoon techniques and ready to execute architecture as "Wham," "Whoosh," or "Clunk."

c. The *Mega-Structure Colonialism* of collective foreign practices that take on emerging markets by developing large structures and drastically altering skylines.

d. The *Primitive Aestheticism* of practices such as those responsible for constructing a circus tent out of Mr. Snuffleupagus in a particular institution's courtyard.[7]

e. The *Parametric Urbanism* of folks who would rather lose their right arm than draw Euclidian shapes.[8]

9. Significance. Star Architects are currently in a position to implode and materialize as new star formations. This is precisely the moment where the Star Architect can be eliminated, and Rising Stars decide either to continue to render architecture in terms of grand narratives or reject the existing model. Equipped with machines, robots, customized tools, collaboratives, participatory ideologies, and a particular consciousness, Rising Stars have the potential to kill off paradigms handed down from their predecessors.

10. Reasoning. The following argument calls for the death of the Star Architect, imploding the supernova and refocusing attention from the image of the Star toward anonymity in architectural productions.

Toward Anonymity
Certain events throw architecture into a state of crisis such as natural disaster, political upheaval, economic recession, or the effects of new technologies within the discipline. These events come from both outside the profession and within, demanding architects to constantly reevaluate their role. As a perpetual milieu, crisis does not conform to a singular problem such as method or style in architecture but is made up of compounding matters. Although all of this crisis-talk sounds bleak, there are a set of opportunistic associations with crisis. A particular "promise of crisis" clearly recognizes the

profession's loss of way or loss of meaning, and in turn ignites curiosity as to how to find it again. Self-reflection of the Star Architect during these times questions an authorial voice while all-encompassing narratives start to become superseded.

To understand present-day predicaments in the profession, it is important to consider the historical crisis that led to the profession of architecture. During the Italian Renaissance, large urban projects erected during the height of demographic and economic expansion were left incomplete.[9] Technical problems discovered in the construction of several cathedrals were associated with events outside of the profession including widespread disease and cultural issues. As a result, this raised question and debate about the role of the architect. These events incited Leon Battista Alberti to write a critical response to Vitruvius's *Ten Books on Architecture*, in the form of his own ten books—an evaluation of architecture's current state in 1452. Alberti's famous treatise, calling for the split of responsibilities between architect, engineer, and craftsman, established what we now know as the profession of architecture and marked the first acknowledgment of the authorial Star Architect.

Alberti not only defined the profession but bestowed colossal authority upon the architect. This new system of ranks in construction situated the Star Architect as an autonomous figure. Territorial roles and clearly defined fields of responsibility eventually led to our current interpretation of the star system and the profession. Most recently, this archaic star model is quietly being reworked as collaborators, makers, programmers, and builders blur the highly territorialized agendas established by Alberti and his intellectual offspring. Both Alberti's territorialization of the profession and the more recent territories claimed by rising Star Architects lack significance among those currently reimaging the production of architecture.

Mario Carpo, architectural historian and critic, proposes a pre-Albertian paradigm by recognizing that "counter to the Albertian principle of separation between notation and construction, digital architects today are increasingly designing and making at the same time."[10] Participatory constructions no longer call for a Star Architect personality, but require collaborative means. Anonymity plays a role in these collectives by working within different processes and yielding new craft forms. Furthermore, a case for digital anonymity is entertained mathematically vis-à-vis the anonymity set, algorithm, or random command—a departure point for many recent projects. The digital project allows for both the possibility of anonymity and the possibility of reuniting the divide between thinker and maker (or designer and craftsperson), as well as realigning architect and architecture.

Questions for the contemporary architect oscillate between repositioning Alberti's split of the professions and reviving Vitruvius's master-builder era. In seeking to position her- or himself in popular society, providing a service that is intelligible, desirable, entertaining, and worth something, the architect is asked to perform the role of a celebrity. This performance distances the architect from architect-as-builder, architect-as-craftsman, and architect-as-service-provider. Although each reading of the architect supports a participatory practice, all are seemingly uncharacteristic of a celebrity. Celebrity and architecture are placed in competition for attention and further separate the architect from architecture.

This argument does not suggest a reversal for the division and separation of roles within the profession. For futures to come, the profession will continue to hold on to a set of clearly separated roles. Yet we might look at the potential in repositioning the architect's value among a more anonymous group. The Star Architect persona with individual brilliance exits, while participatory makers enter the conversation.

Post-Star Paradigm

The examination of two case studies reveals moments within the discipline where anonymity starts to look promising. The Henderson Waves Bridge and the Phaeno Science Center represent two processes in architectural production that eliminate the star paradigm. Operating within a post-star paradigm, both projects relinquish exclusivity in search for architectural value. Emergence of pre-Albertian processes is identified in the Henderson Waves Bridge, a 300-meter pedestrian bridge in Singapore. Collaboration between the architect, engineer, and craftsman allow for technically minded methods to exist in combination with geo-metrical and mathematical strategies. Not only is the collaborative effort noticeable in the develop-ment of the bridge structure, but a value is placed on these processes in the explanation of the archi-tect's practice. When describing the firm's profile, collaborators are mentioned six times within the body of the text. The architect, interested in invent-ing software to maintain rigorous processes, devel-ops custom tools for each project.[11] This allows for the team to carry out demanding tasks with total control of the geometry. Similarly to these ideolo-gies, the pre-Albertian makers of the master-build-er era built machines for construction and small measuring devices, as well as large infrastructure projects.[12] Curiosity in the form of design, craft, engineering, and mathematics calls for discussion among many constituents; all of a sudden, the idea of the star or architect as genius is missing.

Another example where architect, engineer, and craftsman rely on a collaborative model is in the Phaeno Science Center in Wolfsburg, Germany. Here the architect envisioned complex geometries where each architectural element serves a specific purpose; the engineer took a parallel stance where the structure does not spare any redundancy.[13]

The designing and making procedures require full participation from all partners of the team to achieve the final building's structural, programmatic, and aesthetic result.

Systematic pairings between various roles such as architect-engineer, engineer-craftsman, and craftsman-architect are required to execute the total production of architecture. Walls and slabs produce a continuous shell (architect-engineer). German handwerkers form thousands of rebar vectors into a structural meshwork (craftsman-engineer) highly descriptive of the finely calculated geometry (engineer-architect). Self-compacting concrete performs innovative plastic and structural tricks (engineer-craftsman). The project in its entire com-plexity might not exist today if merely designed by the architect and then handed over to the engineer to "build." These shared and transparent processes between architect, engineer, and craftsman were revealed the first time the project was presented to the public. Here, at a lecture, the architect and engineer shared the stage.[14]

Within a post-star paradigm, collaborators are less concerned with understanding the recipe for architectural stardom and more interested in a participatory constituency fully engaged with the production of architecture. By calling for the death of the Star Architect, the argument should not be misunderstood as the removal of authorship or a denial of authenticity from the project. Rather, authors, process, and architectural project become a transparent participation.

Notes

1. For example, Tom Cruise is *Mission Impossible*; the viewer watches the production to see the Star, not the film.
2. For example, Frank O. Gehry is the Guggenheim Museum; the visitor arrives at the building to see the Star Architect, not the architecture.
3. Greg Lynn. "Greg Lynn: Telephone–February 9, 2004," *Perspecta 37: Famous* (Cambridge: MIT Press, 2005), pp. 24–31.
4. Consider the phenomenon of the TED Talks. TED is a nonprofit "devoted to Ideas Worth Spreading." A subtext through many of the presentations is the suggestion that one is both cosmopolitan and part of an avant-garde club by purchasing a limited $7,500 pass to attend the conferences in person.
5. Denise Scott Brown. "Room at the Top? Sexism and the Star System in Architecture," in *Architecture: A Place for Women*, edited by Ellen Perry Berkeley and Matilda McQuaid (London and Washington, DC: Smithsonian Institution Press, 1989), pp. 237–246.
6. Rem Koolhaas. "Rem Koolhaas: The Architect Planning for the Future," CNN Talk Asia, 2009 (http://www.cnn.com/2009/WORLD/asiapcf/06/23/talkasia.remkoolhaas/index.html accessed March 30, 2011).
7. Mr. Snuffleupagus is the Muppet character from the Sesame Street televi-sion program whose figure is formed of thick brown hair. The analogy points toward practices working on primitive piles of material/form and the resulting aesthetic of this work.
8. Patrik Schumacher. "Parametricism and the Autopoiesis of Architecture," lecture, SCI-Arc, Los Angeles, September 13, 2010.
9. Anthony Grafton. *Leon Battista Alberti: Master Builder of the Italian Renaissance* (New York: Hill and Wang, 2000), p. 268.
10. Mario Carpo. *The Alphabet and the Algorithm* (Cambridge, MA: MIT Press, 2011), p. 45.
11. See IJP Architects, www.ijpcorporation.com. Within the practice profile page "About IJP," the architects have listed their collaborators six times when describing the practice's philosophy. Adams Kara Taylor Consulting Civil and Structural Engineers are listed three times.
12. Vitruvius, *The Ten Books on Architecture*, translated by Morris Hicky Morgan (New York: Dover Publications), pp. 283–319.
13. See Adams Kara Taylor Consulting Civil and Structural Engineers, www.akt-uk.com.
14. Hanif Kara, Christos Passas, and Paul Scott, "The Realization of the Phaeno Center, Wolfsburg," lecture, Architectural Association, London, March 2, 2006.

How the Architect Found the Engineer

Svetlana Potapova

In any project, communication among the parties involved can make the difference between a successful and an unsuccessful outcome. In planning, designing, and building a structure, communication between the architect and the engineer is vital at all phases of the project.

A significant part of communicating relies on representing the structure visually. Once the visual representation is achieved, the dialogue that follows depends on the design process and methodology. What are the various design approaches, and what are their differences? Do the differences have an impact on the communication among the parties and, ultimately, an impact on design? There is no such thing as a flawless project delivery, and imperfections often stem from a lack of interaction among the various parties involved in design.

One approach is to view imperfections as constraints and simply develop the best design possible within those constraints. Another approach would be to envision the imperfections as inefficiencies in the process and attempt to mitigate them by inventing something new. Of course exploring new frontiers is always risky, given the pressure of the novelty not working or not being accepted. It seems as if design methodologies and delivery vary based on theamount of risk the parties are willing to accept and their willingness and ability to collaborate.

During the Harvard Graduate School of Design's "In Search of Design Through Engineers" course with Hanif Kara and Andreas Georgoulias, I met my collaborator, Murat Mutlu. We decided to form a partnership and develop tools to facilitate communication between architects and engineers. Here I will document our findings and demonstrate the potential they have in enhancing project delivery. But I would also like to discuss the collaboration process between Murat and me, and relate what I learned from working with an architect. As students, we worked in a risk-free design environment, which is the healthiest for generating creative ideas.

Design Processes and Their Influence on Projects
Murat and I were especially interested in one aspect of the design process: the exchange and manipulation of data that visually defines the structure—a key element of a project. It communicates the concept to a client and provides geometric information to the engineer and contractor for design and construction. The tools used to facilitate this exchange of data are rapidly evolving and worth examining.

Although I was familiar with the way this portion of the design process functioned, most of what I learned had come from class case studies. The in-depth exposure to projects in which a stakeholder took greater risks and involved innovation components allowed me to understand how the project outcome could be manipulated. The design process starts with the architect establishing a concept and representing it as elevation, plan, and section drawings, or through a 3D model. The drawings are then passed to the engineer, who analyzes and designs the structure. Depending on the project, a number of inefficiencies come up at this point. Some can bring the project to a grinding halt, and this is where the idea of risk and innovation returns.

In the case of the Phaeno Center, for example, the building had complicated geometry and thus had to be analyzed as one system in a finite element analysis model. To get a reasonable estimate for the stresses flowing through the structure, a large number of nodes had to be used to describe the structure—more than the model could handle. One way of dealing with this issue would be to tell the architect that the geometry was too complicated and could not be analyzed. Another would be to settle for using fewer nodes and receive a vague solution, and heavily overdesign the structure. Adams Kara Taylor's solution was to improve the software to allow analysis of more nodes. Once the model was working, they were able to further improve the concept by taking out unnecessary material and creating windows. While improving the software, AKT had to deal with the risk of not knowing whether their time investment was worthwhile.

The exposure to such solutions drew my attention to the importance of continuous feedback between the architect and engineer, and how the dialogue enhances the overall project concept. Other types of inefficiencies arise in projects that do not have such a severe impact, but can create delays and annoyances. These usually involve making changes to the models. We were exposed to a few such instances with the Phaeno Center assignment. The architect's model was in no way linked to the finite element analysis model or the concrete analysis model. Every time a change was made to the concept, the engineer's models had to be regenerated.

Parametric geometry definition has emerged as a solution to this problem. When generating a visual representation of the structure parametrically, components are defined in relationship to each other. If a small change is made that affects the whole model, no time is spent on readjusting the remainder of the

model. Furthermore, parametric modeling aims at working with export and import functions to improve the transition between various kinds of software.

Parametric modeling, however, is somewhat inflexible after the constraints and parameters of the system have been defined. Furthermore, parametric geometry definition is not yet available in major structural analysis software packages. After assessing these issues, ideas of computational geometry and design scripting started formulating into concrete plans.

While participating in the "In Search of Design Through Engineers" course, Murat and I were also in a class called "Design Scripting" at the Massachusetts Institute of Technology. The two courses complemented each other beautifully. The Harvard class, particularly the Phaeno Center case study, exposed us to industry tools and their current limitations. The MIT course provided us with a technical platform that we could use to create our own applications.

The field of computational geometry and design scripting involves drawing objects in modeling software using coding language instead of toolbars in user interfaces. The code can be processed as many times as needed, each time regenerating a drawing, making it simple and fast to redraw a structure. The code can also interact with a surface that has already been drawn by the user—for example, code that delineates surfaces can be used to generate meshes, which are then exportable to structural analysis software. In fact, certain architecture and engineering firms have noted the advantages of scripting and are employing specialists in computational geometry to create and improve these tools.

Exposure to the issues presented in the Phaeno Center project gave me a better understanding of how scripting tools can be put to effective use in practice. Murat and I were fascinated with the limitless applications of scripting and wanted to explore them further in our project, while addressing continuous architect-engineer interaction. This led us to exploring form-finding, or shape-finding, for our project.

Form-finding and Project Results

Form-finding is one of the recent buzzwords in the structural engineering and architectural fields. It refers to discovering an optimal state of a structure: structurally efficient and aesthetically pleasing. This is achieved through an iterative process of evaluating the structural stresses and then modifying the structure's geometry to reduce them. This not only reduces the amount of material needed to build the structure, it also smoothes out stress concentrations that could be detrimental to the structure over its lifecycle. Currently no mainstream software exists to perform this process; many firms have been developing in-house programs that do form-finding. Murat and I attempted to create a program that would give an architect a general idea of the stresses in a structure visualized in Rhino, a 3D drawing software typically used for structures with complex curves. The intent was to communicate the state of stress in each member through a color code. The architect could then use this structural analysis feedback and modify the form by either changing the overall geometry or the cross-section of the members.

The structural analysis in the program is based on matrix structural analysis, the underlying strategy for mainstream analysis software; it involves solving large sets of linear equations written as matrices. Originally the goal was to have Rhino solve the equations, but it was discovered that Rhino's matrix operations are very limited; therefore the equations were solved in MATLAB, software that is ideal for manipulations of large sets of data in matrix format. The geometry and connectivity of the structure are defined in Rhino, and an export file is created with vectors of data. This data is then imported into MATLAB, which computes the stresses on the structure under its own weight and exports that information back to Rhino. The information is then used in Rhino to visually communicate the stresses. Murat mainly handled the Rhino code and I took care of the MATLAB portion.

Many challenges were encountered. While working out the method, we had to narrow the scope because of time limitations. The work required to add in 3D geometry and cross-sections of vari-able elements was significant. Thus the program achieves form-finding for 2D frame members with a circular cross-section. This type of application facilitates feedback between engineer and architect, revolutionizing the design process and how architects approach conceptual design.

Murat and I explored many options to our solution and evaluated their advantages and disadvantages. For example, we played with the import-export functions between SAP2000 (mainstream structural analysis software) and Rhino. We discovered that it was difficult to import files that dictated geometry from Rhino into SAP2000 and have them retain the same geometry. Furthermore, cross-section types and loading on the members, if varied, still had to be reassigned one by one. Through the process of exploring many options, we learned an immense amount about the tools in the industry. But this project also became a learning experience regarding collaboration with another discipline.

Lessons Learned

I remember in particular one discussion about the overall concept of form-finding. A shape is selected for a specific load, but a structure—throughout its lifecycle—experiences many loads. The dilemma then becomes selecting the loading for which the form is optimized. While I was describing loads that occur on the structure and how building codes guide engineers through the selection of loads, it dawned on me how imprecise (and lucrative!) the field of live loads (particularly natural loads) definition was. The probability of the structure experiencing the load that it is designed for (a hurricane wind, for example) is low; therefore 99 percent of the time the structure is in a highly inefficient state of overdesign. Furthermore, even though there is a small chance that the structure will experience the design load, it is still not certain that that particular load was estimated properly.

For example, a structure might be designed to withstand Category 4 hurricane winds, but typical storms are in the Category 2 range, so the structure has more stiffness than it needs throughout the majority of its lifecycle. Then a rare hurricane occurs, but instead of a Category 4, it is a Category 5 wind; the structure will probably be destroyed. This can, and has, happened: climate cycles are changing, but the codes have to draw the line somewhere.

With that notion, I presented Murat with some alternatives to controlling live loads that I was studying in my "Motion-Based Design" course at MIT. My professor, Jerome Connor, proposed ideas for controlling the live loads actively. The scheme involves sensing the live load and having the structure respond to it by either generating a reciprocating force through an actuator or varying its stiffness just for that moment in time. This concept is limited by various technologies (sensors, actuators) involved in the design, but they are evolving.

I proposed that the future of fully efficient (utopian, in a sense) structure is one that is form-found for dead load, which is defined with a high degree of certainty and has an active control installed for the live loading. With that, we selected dead load as the loading scenario for which we would shape our structures. Whether the industry is ready for that type of scheme is of course another question.

Renegotiating Roles
It was interesting to see how frequently our roles overlapped as engineer and architect, generating some amazing ideas. While looking at the complex formwork requirements for the columns of the Phaeno Center, for example, Murat pitched the idea of fabric formwork. The Phaeno Center concept was a parametrically derived shape, but the constraints were based purely on aesthetics. I took Murat's idea a step further and proposed that one of the parameters in the derivation of the concept be the fabric formwork, which could be scripted into the code. This not only created an efficient idea; it reinforced our group's concept of revolutionizing the design process—specifically the way architects come up with concepts for the structure.

Murat brought in the idea of formwork, which typically would be associated with the structural engineer, and I raised a notion that is closer to architecture. Trusting Murat to teach me something about engineering but not constrain me to technical subjects promoted a creative and collaborative environment. In fact, I enjoyed learning about engineering through his perspective, particularly when he explained the truss analysis method as he understood it. Whereas I thought of finding the reaction forces as global equilibrium, he simply compared it to a beam with holes—which of course is exactly what it is. This lesson of representing the discipline properly, without assuming ultimate authority on the subject, was valuable to me.

On Knowing Each Other's Work
At one point Murat wanted to know what each portion of the MATLAB code meant; I started explaining it to him, in perhaps unnecessary detail. The project worked well, even though Murat never fully dissected the matrix analysis method and I am not sure how he came up with his searching algorithm to name the members and the nodes. Looking back at the experience, I see that it was not necessary to know how each other's codes worked, but it was necessary to achieve certain levels of communication and knowledge about them. For example, it was vital to communicate to MATLAB how the members get numbered and connected. Therefore it was critical for us to understand the inputs and outputs of our portions of the work, particularly the alternatives that can be achieved and how they would affect the final result.

On Knowing Your Own Work
Murat and I, in different programs, had different schedules; thus the little time we had for meetings had to be optimized. I recall once I had to relate to him the difficulties of numerically calculating moments of inertia for various cross-sections over lunch. I had not used that theory in a number of years; luckily I had the help of a brilliant engineer, Hunter Young, who was sitting next to me. Given our limited time, I realized that I had to be prepared for our meetings for them to be effective. But it was not enough to have the subject make sense in my head. There is a difference between knowing your job so that you can do it and knowing it well enough to

teach it to someone else. Once you cross over into the realm of explaining theories to someone else, you have to recall the underlying assumptions made within that theory and why they were made.

Final Thoughts

The experience of building a form-finding tool with Murat enriched me on many levels. While trying out the applications, we were able to pinpoint areas in need of improvement. Beyond that, we discovered the limitless possibilities that can be inexpensively explored with the design scripting tools that already exist. Although the area of visual data representation and exchange is somewhat limited by technology, it is more limited by the ability to specifically identify the issues that need to be resolved. Identifying the issues can be challenging, requiring willingness to take on risk, a stance of being knowledgeable yet open to learning more about your own field, and commitment to collaboration and more communication time between architects and engineers.

A Necessary Resistance within Architect-Engineer Collaboration

Fai Au

The Evolution of Architects' and Engineers' Roles

The industrial revolution created a new way of looking at buildings and infrastructures. Industrialized steel began to be used in the building of entire structures and new technical universities created a cadre of experts trained to deal with the new materials: iron, steel, reinforced concrete, and pre-stressed concrete.[1] These events led to the emergence of an autonomous profession, engineering, which was previously a branch of architecture.

Despite great late 19th century engineering works such as the Eiffel Tower and the Brooklyn Bridge that demonstrated engineers' creative capability, form givers and city shapers, the role of engineers remained over-shadowed in the mainstream architectural scene. The Bauhaus under Gropius at the time barely recognized the tradition of structural engineering. According to a classic work on the Bauhaus, "Gropius included forty-five illustrations, not one of each shows any work of structural art."[2] Instead, the Bauhaus posited new thinking about technology and design from the perspective of architecture, not structure.

In general, engineering as an equally creative counterpart to architecture is not the case today. The architect is still perceived as a creative individual who shapes our built environment and focuses on aesthetics, while the engineer is viewed as a "technician" who focuses on performance and efficiency, a kind of assistant to architects in solving problems and implementing design concepts. The rise of digital culture in the late 20th century, however drastically altered the educational mode and practice role for architects and engineers, which in turn changed the architect-engineer relationship. Further, the rise of sustainability as a global design issue has required a more collaborative relationship between the two disciplines. With these sophisticated layers of change, how will the architect-engineer relationship evolve in the future? Collaboration between architects and engineers has become a frequently debated topic within the two disciplines. What problems does this collaboration present? How should this collaboration be developed in the future? This essay will examine the fundamental differences between the architecture and engineering professions in terms of education and practice. I will also attempt to propose potential changes in the education and practice system, and finally evoke the idea of resistance, which could possibly be embedded within future architect-engineer collaborations.

The Differences between Architects and Engineers: Education

"Architects are creative, but egotistical, flaky, and self-promoting; Engineers are thorough, but inflexible, stubborn, and socially awkward."[3] If these stereotypes are true to any degree, the causes of such tendencies are worth examining, beginning with a look at the differences between the architect's education and that of the engineer.

Engineering education is structured linearly, providing students with increasing levels of knowledge and skills, and waiting until all of it is absorbed before asking students to "design." By rarely encountering ambiguity and uncertainty, engineering students often assume that problems in practice will have singular solutions and one right answer. With a master's degree in structural engineering, an engineer can analyze and size a predetermined beam or truss without the need to design them. From the initial learning stage, the idea of being a problem solver is embedded in the mindset of engineering students. The studio structure of architecture education, on the other hand, creates a different learning environment. Students often encounter problems that are ambiguous and uncertain. There are rarely right answers to their problems, only better or worse proposals. Students are encouraged to be creative and imaginative.[4] Architects and engineers are thus trained to position themselves in opposite roles.

Recent Developments in Architectural and Engineering Education
Since the rise of digital culture during the late 20th century, computers have played an increasingly essential role as the analysis tool for structural engineering. The graphical method—the only way to design masonry and bridges in the 19th and early 20th century—has been largely abandoned.[5] When the subject is presented through complex mathematical exercises on computers, without the use of visualization and approximation methods, engineering students tend to lose their intuitive understanding of structural behavior and are no longer trained to simplify problems visually. They become expert in analysis, but with little fostering of their creative imagination.[6] Developments in computation technology have also provided design tools for architecture students. Yet instead of abandoning the visual, architecture students are becoming even more capable of visualizing and representing design with three-dimensional modeling tools. It is not surprising to see students in architecture schools using elaborate analytical diagrams to illustrate sophisticated layering and assembly of programs and spaces. The essence of this obsession with visualizing design analysis and the exuberant use of diagrams may not be the factual representation of the performance and efficiency of the design, but rather relates to the power of images.[7] Instead of being factual, these diagrams and images produced by computers are perhaps more "fictional" and even "ornamental" in the realm of architectural representation.

The introduction of computation technology has changed architecture and engineering education drastically. Instead of bringing the two types of education closer, the technology seems to widen the gap: engineering students get more deeply involved with complex mathematical analysis, without thinking visually like architects; architecture students become focused on the visual, which does not address factual and performance aspects.

The Differences between Architects and Engineers: Practice
In general, when engineering students and architecture students leave school, they have already been inculcated with different career perspectives. Engineering students are trained almost from the start to be specialists, and join a large office. They are trained to do work of focused scope. They are heavily exposed to mathematics and hence to axioms, which cannot often be challenged. In other words, an average engineering student learns not to challenge. Conversely, architecture students are trained to be more independent. They are told, "You will become a great architect and you should feel free to behave as one." They are trained to challenge their critics and be proud of their architectural conceptions.[8]

When graduates with such different perspectives enter practice and work together on a project, they may clash. When the "great" architect meets with the "modest" engineer, the latter does not dare to contradict the architect's requests and is often compelled to find unjustifiable solutions for structures. Or the "great" engineer may meet with the "not-yet-great" architect and authoritatively deny him the realization of his dream. In the majority of the cases, average architects meet with the average engineers; when professionals with such different approaches to the same problem have equal standing, a greater clash may be expected.[9]

Recent Developments in Architectural and Engineering Practices
Today the two professions are drawn together to collaborate, driven by multiple forces that have emerged since the late 20th century. First, there is a great increase in complexity in the building industry.

Billington defined the separate roles of architects and engineers in the past:

> Structural designers give the form to objects that are of relatively large scale and of single use... Architectural designers, on the other hand, give form to objects that of relatively small scale and of complex human use... the prototypical engineering form–the public bridge–required no architect, the prototypical architectural form–the private house–requires no engineer... the two types of designers act predominantly in different spheres.[10]

It is almost impossible to put the two disciplines in different spheres today. With the complex amalgamation of urban infrastructure, civil engineering, and building design at various scales, together with the sophisticated juxtaposition of information networks and utility systems, the tasks for architects and engineers are intertwined. No matter how different their approaches to the same problem, architects and engineers are pushed to work together.

The second force is the rise of sustainability as a global design issue. Building codes that relate to environmental performance and rating systems such as LEED develop worldwide and become more stringent. With such an increase in demand for technical assessments and analysis, architects have to collaborate with environmental engineers to fulfill the new requirements. However, architects do not always engage in elaborate environmental analysis design. They rather welcome the incorporation of prescribed sustainable design elements. Sky gardens on the body of super towers, symbolic green fields wrapping on rooftop, and computer controlled shading devices on facades appear frequently on numerous competitions and projects. This can be an indication of limited incorporation of environmental design, leaving space for improvement in the collaboration between architects and environmental engineers.

The third force is the fundamental change of the architect's perception of structure and his or hers relationship to it. This change occurred in the last quarter of the previous century, with the advent of high-tech architecture. The structures of high-tech architecture exploit technological progress in the development of materials and in production techniques. Their popularity may be traced to a trend in which the flow of forces is made visible. The outcome revealed itself "in a quality of lightness and transparency: compression forces flow through tubular members, tension forces through rods or cables."[11] This approach was welcomed by engineers and offered them an opportunity to make original contributions. It was the moment at which the interests of architects and engineers converged. The engineers did not notice that they were exaggerating and displaying the flow of forces with as many nodes, fittings and joints as possible- the same exaggeration the architects desired. This phenomenon was "a form of exhibitionism: elitist high-tech thinking is turned into high-effect showmanship."[12]

From High-Tech to High-Effect

This "high-effect showmanship" did not fade away. Today architects and engineers are working even more collaboratively to achieve this effect. This trend is mostly apparent in developing cities in the Middle East and China, where money is secondary to prestige and self-representation. Architects have shown a strong tendency in exhibiting structural elements on building facades. To achieve this, architects and structural engineers collaborate at the early concept stage. However, the focus of this early collaboration is not merely on performance and efficiency aspects, but also on aesthetics and affects. Compared to the structures of high-tech architecture, the structural elements foregrounded serve much more like diagrams and ornaments that symbolize the idea of structure, instead of expressing the physical presence of structure. The facades of Rem Koolhaas's CCTV Tower in Beijing revealed the designer's clear intension of making apparent the structural elements. The dimensions and density of the steel cross-bracing at different locations of the building façades vary in response to local stress. The patterns on the façade are the literal expression of structural analysis diagrams. However, the cross-bracing revealed on the exterior is not the only structural element supporting the façade; another vertical structural system on the periphery of the tower is completely hidden behind the reflective curtain walls. In fact, multiple layers of architectural details juxtapose in a complex way to highlight traces of the cross-bracing.

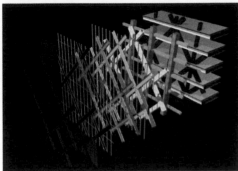

CCTV façade assembly diagram

CCTV Tower Façade

The design effort was spent on revealing the "images" of structure, instead of the structure itself. These "images" of structures are selectively expressed not totally representative of the structure underneath. Some of the structural traces appear broken on the building façade, which was not in the early computer rendering.

Another recent example is Rafael Moneo's new building at Columbia University. Unlike most of Moneo's previous works, which are infused with the sensibility of context, material tectonic, and space, this project shows the architect's desire to visualize the trace of structural force on its surface. Customized aluminum louver panels engraved with patterns, identical to the geometry of the structural trusses, are used to cover the structural frames underneath. The physical presence of the structure is hidden behind these prefabricated panels, and the idea of structure is revealed on the surfaces. For both the CCTV tower and the Columbia building, the design strategies related to the visualization of structural elements unrelated to cost and efficiency. If cost and efficiency were concerns, simply installing a curtain wall system with conventional panels on top of the structural frame of the CCTV tower would have reduced the complexity of finishing details and minimized the wastage in the production of nonstandard glass panels. Likewise, simply installing standardized metal panels to cover the façades of the Columbia building instead of using customized louver panels would have reduced the manufacturing cost. When architects' perception of structure changed and there emerged a desire to show structural traces, some engineers welcomed this interest and were willing to assist architects in exploring it. They

tend to place "elegance" in an equal position with "economy" and "efficiency."[13]

Another interesting development is that since the twentieth century, many architects have been fascinated by using scientific knowledge as a way to approach design problems and develop concepts.[14] Very often this fascination with the sciences is made manifest without the participation of engineers, who usually centralize the scientific dimension as their primary criterion. This could explain why the majority of parametric design projects today remain small in scale, as most were designed with computational calculation that does not relate to structural analysis, and no engineer was involved in the conceptual design stage. It would be challenging for large-scale parametric design projects to be realized when rationalization in terms of structural feasibility comes in at a late design stage.

The Henderson Waves Bridge by George Legendre was designed with the engineering software MathCad, but without the involvement of engineers at the concept stage. Through manipulating mathematical formulas, the architect managed to generate an elegant wavy form of steel ribs along the path of the bridge platform. Although the steel ribs suggest a structure that holds up the bridge, their curvilinear form runs along a different path compared to the moment distribution.[15] In fact, only about one-fifth of the steel ribs are structural elements; the others serve to retain the completeness of the architectural form. This is an example that shows how architects' fascination with science and structure could be totally irrelevant to the collaboration with engineers.

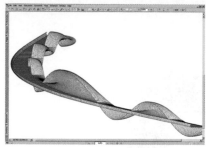

Henderson Waves Bridge

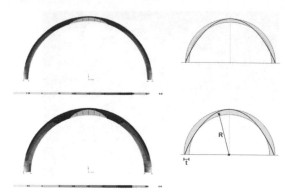

Comparison of finite element and limit state analyses

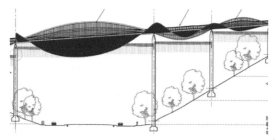

Henderson Waves Bridge: structural analysis

Filling the Gap in Education

The gap in the two education systems should be filled. In any conventional building project, where even minimal collaboration between architects and engineers is required, basic communication is expected. If architecture and engineering students are trained to have such different mindsets in school, they cannot easily communicate. The introduction of advanced computer design software in architecture and engineering schools has even pulled the two further apart in terms of design methodology: engineers become experts in analysis, without fostering their creative imagination; architects become experts in visual representation, without a solid understanding of structural principles. It is worth considering the reintroduction of graphic analysis methods into engineering education, along with three-dimensional drawing. Without these skills, engineers have difficulty in both imagining the complex interrelationships of a building and presenting their ideas. Some academic research has illustrated that the graphic analysis method could provide a more intuitive and accurate check of structural stability than computer analysis methods.[16]

It is also worth considering adding more structures classes to design studios in architecture school. Although most architecture schools offer structures courses, they tend to remain at a generic level, and it is rare to see structural analysis incorporated into the design studio programs. If architecture students could be trained to use the graphic method of analysis to investigate structural feasibility, they could have more opportunities to explore the possibilities for structure as an aesthetic generator and form finder.[17]

The "Star" System

In the "star system" of architects, practitioners such as Rem Koolhaas and Jacques Herzog appear as heroes who garner public attention, and they often inspire young architects to be ambitious and enthusiastic about their future careers. However, "star" almost does not exist in the engineering profession today. Cecil Balmond is one of few exceptions, having actively collaborated with star architects and extended his contribution into the architecture field. In his publications, Balmond discusses ideas such as patterns, hybrids, rhythm, etc.–concepts usually raised only by architects.[18] Although Balmond's position could not represent the majority of the engineering profession and does not suggest a direction for the majority of engineers, the "star" effect he created does contribute to better public recognition of engineering. It also gives incentives for young engineers to reestablish their ambition to have a stronger position in society. Perhaps it is worth encouraging young engineers who are more adventurous and open-minded, and have a strong interest in architecture to move more decisively toward investigation and discussion of architecture. When engineers gain more self-satisfaction from being creative counterparts to architects, more synchronized dialogues could happen. Although the star system has been condemned from many perspectives, the star effect could help the engineering profession become more vibrant and ambitious.

The Combined Force of Young Professionals
Architect-engineer combined practices represent a positive way for architects and engineers to strive for excellence under the same roof. They provide opportunities for the two disciplines to have daily communication about design and construction, in terms of not only aesthetics but also performance and efficiency. Moreover, joint practices also diversify the scope of services, which reinforce the firm's brand and economic performance.

Although it would be difficult for large firms to accommodate this change because of deeply rooted cultures and organizational structures, young architects and engineers about to start their own firms should consider the possibility of joint practice. There are many examples of combined practices of architects and other design professionals, but groupings of architects and engineers are limited. If the recommended changes could bring the two disciplines closer, it would be easier for young architects and engineers to seek opportunities for joint practice.

Resistance within Collaboration
In reaction to the ambiguity of architect-engineer collaboration today, I would like to apply the term "resistance," which should be embedded within the idea of collaboration. "Resistance" does not mean antagonistic confrontation, but its more subtle meaning comes in layers. First, resistance means the opposition to the freedom that architects and engineers have today. With highly advanced computation tools, architects have tremendous freedom to produce complex forms in a rapid manner. In most cases, the architectural imagination operates without thorough consideration of structural principles or the participation of engineers. Were engineers involved

in the mathematical calculation process of George Legendre's Henderson Waves Bridge, the design process would have been more closely related to structural analysis and may have minimized structural redundancy whilst maintaining the formal elegance of the steel ribs. Similarly, engineers are "free" from the graphic analysis method because of the availability of advanced structural analysis software. However, they are participating less in the creative process of generating three dimensional forms. For both disciplines, perhaps it is worth considering a certain amount of resistance to this freedom to facilitate productive collaboration.

The second layer of resistance is resistance of exuberance, a concept often applauded by architects today. Peter Cook, for example, urges us to recognize opportunities to let go creatively, "in a rush of torrent, unencumbered by pace, taste, hierarchy, chosen moment arguable relevance, correctness and sustainability credentials."[19] Seen from a positive perspective, this article toward exuberance yields creativity and imagination; from a negative perspective, design exuberance, especially with its realization in built form, tends to emphasize aesthetics and elegance more than performance and efficiency. The exuberant exhibitionism of structures (or ornaments?) of Koolhaas's CCTV tower, Herzog and de Meuron's Beijing National Stadium, and Dubai's Burj Khalifa tower blend the political and economic desire to build with architects' and engineers' obsession with complexity and showmanship. Even sustainable design leans toward exuberance. Complex and intelligent systems are designed and installed in buildings to achieve the energy-saving performance. To speak of "energy-saving buildings" ignores the fact that the energy required for the manufacture and operation of the technical systems may far exceed the savings from passive or active use of solar energy. Both architects and engineers are responsible for this "fashionable excesses"—they maximize areas of glazing regardless of the location and orientation of the building, and then install sun shading and protection against glare, using these as a convenient excuse for the specification of double-skin façades.[20] Perhaps when architects and engineers collaborate and work either on structural design or sustainability design, they should exhibit a certain level of resistance to exuberance. This could rescue designers from the exhausted production of forms and minimize consumption of natural resources and building materials.

The third layer of resistance is the resistance to sameness. A highly promoted idea today in the building industry is the ultimate unification of different professions, systems, and software: the concept of integrated project delivery aims at unifying the various project actors into one building team. A return to the master builder concept, the development of building information modeling aims at unifying all drawing software into a single system and encourages all building professions to work within the same platform. There is no doubt that this unification will better coordinate work and enable efficient communication. However, differences in viewpoints and a certain level of tension among disciplines could catalyze debates and ensure a self-critical attitude.

When every member of the building team shares the same responsibilities and risks, roles tend to be homogenized. This is why resistance to sameness should be sought. With resistance to sameness, architects should be fascinated by engineers' pursuit of facts, performance, and efficiency, while positioning their own design investigations relative to history, culture, and society. Engineers should not be afraid of investigating the poetry and elegance of architectural form and geometry, while searching for precision and perfection in structural design.

In conclusion, architects and engineers should exhibit certain level of resistance, or intermission of play, throughout their collaborative process. Through the silence of intermission, each party can position themselves from a distance in order to retrospectively analyze their performance. Resistance in architect-engineer collaboration is helpful to reflect on and define relative positions, before crossing the borders and boundaries that demarcate each discipline.

Notes

1. D. Gans (ed.). *Bridging the Gap: Rethinking the Relationship of Architect and Engineer* (New York: Van Nostrand Reinhold, 1991), pp. 2-4.
2. D. P. Billington. *The Tower and the bridge. The New art of structural engineering*, 1983. New edition (Princeton: Princeton University Press, 1985) "A New Tradition: Art in Engineering," p. 7.
3. C. F. Blozies. "Square Wheels or Round? – Professional Relationships in Transition" in *ArcCA, the journal of the American Institute of Architects*, California Council (2008, n.3) p.11.
4. K. Simonen. "Engineers are Square, Architects are Spiral" in *ArcCA, the journal of the American Institute of Architects*, California Council (2008, n.3) p.15.
5. For a detailed research in the graphical method, please refer to: P. Block. *Equilibrium systems: Studies in Masonry Structure*, Thesis for Master of Science in Architecture Studies at the Massachusetts Institute of Technology, 2005, pp. 2-12.
6. K. Simonen. "Engineers are Square, Architects are Spiral", p.16.
7. For more details on the power of images, please see: A. Picon. "Architecture and the Sciences: Scientific Accuracy or Productive Misunderstanding?", in Akos Moravanszky, Ole W. Fischer (eds.). *Precisions: Architecture between Sciences and the Arts* (Berlin: Jovis, 2008), pp. 63-71.
8. D. Gans (ed). *Bridging the Gap: Rethinking the Relationship of Architect and Engineer* (New York: Van Nostrand Reinhold, 1991), pp. xiv –xv.
9. D. Gans (ed). *Bridging the Gap: Rethinking the Relationship of Architect and Engineer*, pp. xiv–xv.
10. D. P. Billington. *The Tower and the bridge. The New art of structural engineering*, 1983. New edition (Princeton: Princeton University Press, 1985) "A New Tradition: Art in Engineering," p. 7.
11. J. Schlaich. "Engineer and Architect" in *Detail* (English ed.) (January-February, 2006 v.1) p.7.

12. J. Schlaich. "Engineer and Architect", p.7.
13. Billington in his article "The Tower and the bridge. The New art of structural engineering" pp.16-17, points out the three dimensions of structure: Efficiency, the scientific dimension; Economy, the social dimension; Elegance, the symbolic dimension. He believes that the scientific criterion is primary for the structural designer, yet efficiency must balance with the other two dimensions.
14. A.Picon. "Architecture and the Sciences: Scientific Accuracy or Productive Misunderstanding?", in Akos Moravanszky, Ole W. Fischer (eds.). *Precisions: Architecture between Sciences and the Arts* (Berlin: Jovis, 2008), p. 55.
15. The structural analysis of the Henderson Crossing project was done in the class GSD 6328 "In Search of Design Through Engineer" by Azadeh Omidfar, Evangelos Kotsioris, Fai Au, Fernando Pereira Mosqueira, Xiao Yin, dated 22 Feb 2010.
16. Philip Block, in his MIT Thesis *Equilibrium systems: Studies in Masonry Structure*, pp.9-10, demonstrates that the finite element analysis gives an unsafe and deceptive result about the structural calculation of arches, without showing anything about the stability or potential collapse of an arch. A thrust line gives more information about collapse than a FEA and provides an immediate check for the stability of unreinforced masonry structures.
17. K. Simonen. "Engineers are Square, Architects are Spiral" in *ArcCA, the journal of the American Institute of Architects*, California Council (2008, n.3) p.17.
18. For Balmond's theory, please see: N. Yoshida. "Cecil Balmond" in *Architecture and Urbanism* (November 2006, Special Issue).
19. P. Cook. "The New Delfina" in *Architectural design* (March-April 2010, v.80, n.2) p.58-59.
20. J. Schlaich. "Engineer and Architect" in *Detail* (January-February, 2006 v.1) p.8.

Interdisciplinary Collaboration: Enabling Architects to Regain Leadership in the Building Industry

Advait M. Sambhare

The enthusiasm and optimism with which I graduated from architecture college in 2005 was soon followed by despair and disillusionment, after the sad realization of the limited ability of practicing architects to affect the condition of the built environment. Their traditional roles within the building industry and as the client's trusted advisor were being gradually lost to a newer breed of project management consultants. My anguish was exacerbated by the ambiguity in the mind of the general public regarding the role of architects, who were increasingly seen as mere stylists–a highbrow, artistic accompaniment to the real craft of building that was carried out by builders and engineers. According to a survey conducted by the Construction Managers Association of America in 2005, a staggering 92 percent of owners complained that architects' drawings were insufficient to construct a building.

In an article aptly titled "Can This Profession Be Saved?" Thomas Fisher wrote in 1994 that architects believed that the profession would undergo a fundamental transformation, with a need to revisit established methods of working.[1] Fisher, writing in the middle of a recession, argued that the economic downturn was not the origin of the questions being asked of the profession, though it intensified those issues and brought a sense of urgency to the need to address them. Although new technology and numerous specializations have indeed altered processes in practice, it can be argued that there has been no fundamental change in our conception of architectural practice.

The Progressive Erosion of Architects' Position in Recent History

Even after architects lost their status as master builders when the professions of architecture and engineering split in the early nineteenth century, the position of master coordinator among the various disciplines serving the built environment still earned the architect authority and respect. Since then, however, with the growing scale and complexity of building projects, other professionals have claimed some of the architect's traditional scope, leaving merely the aesthetic aspect, which came to be seen—and with it perhaps the architect too as increasingly superfluous.

While technology has helped expedite design and production processes, it has also led to increasing commoditization in the industry. Despite new technologies, architects are largely unable to assure the performance of buildings and have progressively distanced themselves from issues of cost and constructability, resulting in limited effectiveness. The rift between academia and practice has grown over several decades, severely hindering the ability of graduates to grapple with a changed work environment. Recent trends, such as the development of simulation technologies and digital fabrication, are steps in the right direction, but have yet to find wide acceptance.

The delineation between the roles and responsibilities of architects and engineers has slowly become a deep schism, particularly with the rise of professional liability. Contracts began to get more convoluted and defensive in safeguarding the interests of individual parties, resulting in a growing atmosphere of mistrust between the two professions, as well as toward the contractor or builder. Consequently, owners and clients became increasingly disgruntled with professionals, and in particular with architects, who were traditionally expected to lead the project. This led to the emergence of independent project managers and client representatives, adding another layer of complication to the project. As the architect's credibility waned and his or her scope reduced, the benefit was reaped in full measure by lawyers who, especially in the United States, thrived on exploiting the atmosphere of defensiveness and suspicion prevalent among members of a project team. According to Hanif Kara, "The seams, borders, and divisions (between architecture and engineering) are so clear. These gaps are where the lawyers work."[2]

Leadership by Design: The Architect's Value Proposition and Its Significance
In thinking about what the architect's value proposition may be, it is important to consider the aspect of core competence. Gary Hamel and C.K. Prahalad point out that a clarification of core competencies enables organizations to support and leverage their competitive advantage by aligning strategy and processes accordingly.[3] The architect's core competence is design, which is a powerful capability, because it facilitates coordination among numerous disparate skills to suit the unique requirement of each project. Having a global view of the entire project process enables the architect to act as a trusted advisor to the client.

Although architects would still like the prestige and influence of that position, they haven't been able to adapt their mindsets and working methods to meet the vastly expanded set of challenges and expectations for that role in the present day. Meanwhile, another important dimension of the architect's unique global view further establishes the significance of his or her leadership role: architects do not just merely serve the client's interest but also have an obligation to society and to the environment, as outlined in the AIA Code of Ethics and Professional Conduct.[4]

Reclaiming Lost Ground: Bridging the Architect-Engineer Divide
The historic position of the master builder in the Vitruvian sense, as one who seamlessly and simultaneously pursues the goals of firmness, utility, and beauty, is difficult to reclaim. But perhaps doing so is unnecessary, and even irrelevant, given the scale and complexity of projects today. The new key is in recognizing the engineer and the constructor as valuable strategic allies who can help the architect overcome technical deficiencies and produce more successful outcomes from design. One prevalent trend has been the emergence of multidisciplinary behemoths, in the form of gigantic construction companies with design and engineering functions, major engineering firms that have in-house design capabilities, and big architecture/engineering (A/E) firms. The last recession, however, prompted at least part of the industry to move away from large, multidisciplinary firms and toward the other end of the spectrum in terms of ideology and strategy—the highly specialized but inherently flexible and collaborative entity.[5] Many such design-led collaboratives have emerged with great success, with a focus on being progressively better rather than simply bigger.

Is interdisciplinary collaboration a counterpoint to multidisciplinary integration—and perhaps the way forward for the construction industry? A comparison of these two markedly different approaches to practice is worth consideration.

Multidisciplinary Integration: A Twentieth-Century Perspective
Both multidisciplinary integration and interdisciplinary collaboration are essentially rooted in the same premise. From a twentieth-century contextual view, Walter Gropius's vision of "a closely cooperating team" consisting of the architect, engineer, and constructor, working toward a "fusion of art, science, and business" resulted in the founding of The Architects' Collaborative (TAC) in Cambridge, Massachusetts, where he taught at the Graduate School of Design at Harvard.[6]

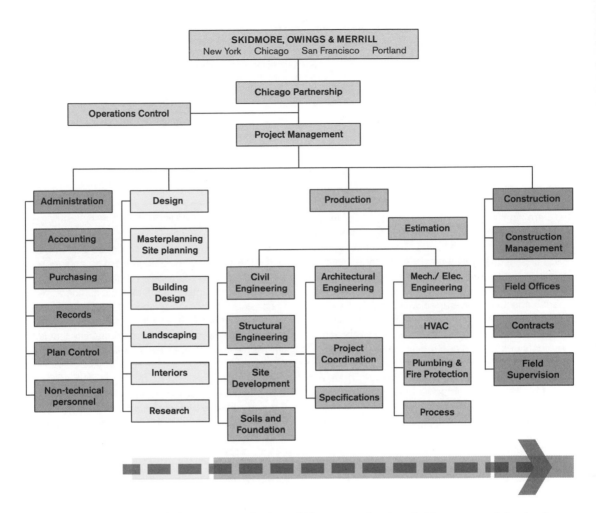

SKIDMORE, OWINGS & MERRILL
New York   Chicago   San Francisco   Portland

Chicago Partnership

Operations Control

Project Management

Administration
- Accounting
- Purchasing
- Records
- Plan Control
- Non-technical personnel

Design
- Masterplanning Site planning
- Building Design
- Landscaping
- Interiors
- Research

Production
- Estimation
- Civil Engineering
- Structural Engineering
- Site Development
- Soils and Foundation
- Architectural Engineering
- Project Coordination
- Specifications
- Mech./ Elec. Engineering
- HVAC
- Plumbing & Fire Protection
- Process

Construction
- Construction Management
- Field Offices
- Contracts
- Field Supervision

-- stratified organization - horizontal (departments) and vertical (management structure)

-- linear, 'assembly line' mode of working, privileging efficiency over excellence

-- management privileged over design, where the latter
becomes a 'department' rather than a driver

SOM organization diagram, 1957

The early twentieth-century idea for integration was taken to a larger scale by Skidmore, Owings and Merrill, LLP, for example. While the first two named partners in SOM were architects, the third was an engineer. Their ambition for "greater scope, greater volume, and greater power to change people's minds" fueled their stupendous growth to include more than 1,000 employees as early as 1958.[7] However, Bernard Michael Boyle critiques that approach as not being truly collaborative in the way that Gropius espoused, but more akin to an assembly line, where particular people were assigned particular roles.

There are significant limitations to the mode of multidisciplinary integration, particularly in the context of large firms, which I argue have a direct consequence for the design quality of their work. The large, multidisciplinary firms are primarily resource driven. There is a need to keep all departments busy and have their capacity adequately leveraged, which is a challenge given that the ancillary services might experience inconsistent demand. This has a bearing on the kinds of projects undertaken and the processes through which they are delivered. Since there are significant economies of scale in this mode of working, there is a natural tendency (even imperative) to get bigger.

In the case of architecture/engineering firms, there is a shift in allocation of resources away from the core competence of design, and in engineering/architecture firms, where design is an ancillary service, it never receives the same focus as the engineering function in terms of capability development, etc. This condition is greatly exacerbated during an economic downturn, straining the firm's resources and hampering its ability to adapt to changing requirements. In addition, many of the large integrated companies are publicly listed, and the pressure of meeting periodic financial performance targets puts further pressure on the design output.

All of the above has a direct bearing on the kind of talent that these firms attract. Often, talented structural engineers would rather work with a top-notch engineering consultancy than play a supporting role within an A/E firm. Similarly, talented designers often prefer to work for an architecture firm with a

Multidisciplinary Integration	Interdisciplinary Collaboration
Tend to be large, resource-driven organizations to optimize economies of scale.	Tend to be small or midsized, culture-driven.
Tend to be large, stratified organizations that may hinder talent development and learning; less likely to attract best talent.	Close-knit organizations produce favorable environment for talent development and learning; more likely to attract best talent.
Focus from design diverted to varying extent. Since demand for ancillary services is unpredictable, may be stressful during downturns in particular markets or in industry overall.	Strong focus on design as core competence. Greater investment of resources in enhancement and leverage of core competence.
Client care is a challenge given the scale of operations, especially when multiple services are engaged.	Clients have greater confidence given the direct involvement of and relationship with firm partners.
Likely to have vested interest in selling services of as many different disciplines within organization as possible, to increase its scope on a project.	Best team made up of most suitable agencies, depending on particular skills and expertise to meet unique requirements of a given project.

Comparative advantages of interdisciplinary collaboration over multidisciplinary integration

strong reputation for design excellence than operate within the architecture component of an engineering or construction company. These limitations of the multidisciplinary integrated firm make it imperative to consider an alternative mode of practice. It is in the pursuit of design distinction without compromise of technical soundness and constructability that is driving the reemergent trend of interdisciplinary collaboration.

Interdisciplinary Collaboration: Leveraging Core Competencies to Deliver Design Excellence
Collaboration is hardly new. Guy Nordenson contextualizes its growing incidence by reminding us that in history, architect-engineer collaborations arose "as a backlash against the generation of architects exemplified by the postmodernists that refused technical input on their designs." He recalls the early twentieth-century precedents of close working partnerships between architects and engineers such as Gordon Bunshaft and Marcel Breuer with Paul Weidlinger and Saarinen with Fred Severud.[8]

The earlier example of H.H. Richardson and O.W. Norcross is also noteworthy.[9] The duo worked together on many important commissions including Sever Hall in Harvard Yard and Trinity Church in Copley Square in Boston. They were not engaged in a design-build contractual arrangement as we know it today; their partnership was defined by a professional understanding and a commitment to the project, rather than a legal instrument.

| RESEARCH
Housing Prototype
Study | → | SITE SELECTION | → | DETAIL DESIGN,
TENDER | → | CONSTRUCTION |

- Investment in speculative project

- Parallel proposals by Laing O'Rourke and Bovis, with AHMM and AKT appointed by client as a shared resource to both teams

- Bovis picked over Laing, appropriates work of AHMM, AKT, and Laing from QMWC. Client gets best deal, but questions of intellectual property in collaborative work arise.

- Contractor in charge, consultants novated to them. Cost and schedule become primary concerns

- Challenge for design to respond to these constraints. Engineer becomes ally to the architect in enabling design to embrace the challenge.

- Brownfield site remediation threatens to usurp major chunk of budget, successfully dealt with.

- Once site is selected, client in hurry, desires completion as fast as possible and prefers not to get involved in daily working, leaving very little time for detail design.

- AHMM and AKT concurrently work on 5 bid packages

- For ease of procurement, risk management, and speed of construction, balconies made part of facade bid. Design and detailing has to respond to this requirement

- Desire to enhance speed of construction results in decision to prefabricate - unitized facade panels, bathroom pods.

- Tunnel-form construction

- Prefabricaed reinforcement - roll out on site like a carpet

Adelaide Wharf: Key Issues

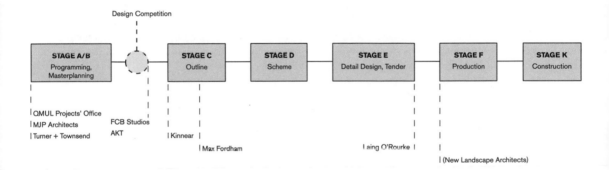

Design Competition

| STAGE A/B
Programming,
Masterplanning | ○ | STAGE C
Outline | STAGE D
Scheme | STAGE E
Detail Design, Tender | STAGE F
Production | STAGE K
Construction |

QMUL Projects' Office
MJP Architects
Turner + Townsend

FCB Studios
AKT

Kinnear

Max Fordham

Laing O'Rourke

(New Landscape Architects)

- Brownfield site limitations
  Soil conditions, Foundation decisions

- Participation of architects and engineers in the business case of the project

- Cost and time saving discussion
  Decisions to prefabricate bathroom pods

- Service engineer not a participant from the very beginning; potential gaps

- Delivery method changed to Design-build, changes project dynamic, ostensibly to lower owner's risk, but a need arises for revisiting many previous decisions

- Political dimension - contractor seeks greater influence within project; replaces landscape architect. AKT survives.

- Balconies detail becomes not just an aesthetic but also procurement issue

- Tunnel form decision

Contractor chooses to execute building facade in masonry; large lag between completition of tunnel form structure and basic facade

Note: The above representation is indicative and based on the available information only

Queen Mary Westfield Students' Village: Key Issues

To counter the increasing commoditization of architectural services, firms must differentiate themselves from competitors. In a recent article, Andrew Pressman urged architects to take advantage of the bleak economic climate to develop collaborative working skills, to enhance their capability to effectively and efficiently coordinate the delivery of complex projects.[10] An interdisciplinary collaborative approach enables firms to focus on their core competencies. When the leadership of small or mid-sized firms is concerned about a collaborative culture, everyone can be involved more directly in the running of projects, and hence with clients. A stronger focus on both client and process enables better design, reinforcing the culture of collaboration as well as positively affecting talent retention.

Striking a balance between focusing on core competencies and orchestrating the coordination of various technical inputs on complex projects is a challenge, and this is where interdisciplinary collaboration is most vital, as shown by firms such as Allford Hall Monaghan Morris. A study of their Adelaide Wharf housing project in Hackney, London, offers a noteworthy example. Projects such as these often depend on funding, regulatory provisions, and political climate. The early involvement and close work between architects and structural engineers enabled more informed design proposals that were technically sound and cost considerate even at the schematic stage, facilitating quick progress from speculative research on prototypes to the actual project. Moreover, the architect and engineer working in close coordination resulted in considerable design value being realized despite stringent cost constraints. In fact, the architects were able to approach issues of design, cost, constructability, and procurement in a holistic way at various stages in the project's evolution.

As firms try to diversify their project portfolio to make themselves more resilient to economic downturns, a collaborative approach enables faster results than trying to develop all of the required skills internally. Instead, as Andrew Pressman states, a collaborative approach can enable firms to overcome limitations in geographic reach and expertise relatively easily.

In the context of the automobile industry, Jay Lorsch and Thomas Tierney observe that backward integration was found to be expensive and ineffective, while a collaborative network of specialized providers enabled better results, more effective resource management, and a safeguard against inconsistent demand.[11] This can easily be applied to architecture firms as well, with the emphasis on coordinating a diverse range of disciplines in a collaborative manner, rather than internalizing many of those functions within a single person or firm.

The key to successful interdisciplinary collaboration is in understanding that it is not a technology but rather a psychology.[12] Collaboration is not a process that can be codified into a set system; it is more of an attitude that needs to be inculcated in the culture of a firm. It begins with every participant acknowledging that each of the others brings something valuable to the project and that their combined intelligence is more likely to deliver positive results than working in isolated silos. This can be challenging for architects, since a culture of pride in individual authorship is deeply ingrained in the profession.[13]

A culture of collaboration is important to the individuals within a project team and even more critical between teams. As mentioned earlier, external factors have a huge impact on the success or failure of projects. As Feilden Clegg Bradley Studios demonstrated in their Queen Mary Westfield Students' Village project, getting involved in the business case of the project assures the clients that the architect understands their priorities. A close collaborative relationship with AKT, the structural engineers, safeguarded both sides from falling prey to changing project dimensions (in this case, the technical team became novated to the contractor, who began wrestling for greater control by attempting to bring in a new team of professionals). Interdisciplinary collaboration ensures alignment of interests among the various constituents of the project team, thereby reducing the risk that changing project dynamics could hamper a successful outcome.

The fallacy of the integrated firm lies in the fact that although a firm may have all specialties within it, the multitude of skills aren't present in every individual, so in effect, they too are relying on a collaborative approach behind their office doors. Because the integrated entities tend to be large and resource driven, efficiency takes precedence over other aspects. Such firms have a vested interest in selling services to a client, from as many of its departments as they can, regardless of whether they are best suited to the job. This is where the biggest advantage of the interdisciplinary collaborative firm comes to the fore: on every project, no matter the nuanced requirements, a team of experts can be assembled to deliver specialized services, increasing the likelihood of successful outcomes. This is also reflected in Hamel and Prahalad's view that "forging strategic alliances" is one of the key strategies in leveraging an organization's core competencies to gain competitive advantage in the marketplace.[14]

The move toward a more collaborative mode of working is gaining momentum. Technology and the delivery method of projects are becoming catalysts for this change. The increasing sophistication of building information modeling is enabling collaborative work between professionals to be taken to the next level, but fostering a psychology of collaboration is more important than any technology. The project must be brought back to the center. Integrated project delivery is gaining currency in the United States as a possible counter to the mistrust arising from rigid contracts within the conventional project structure, illustrating a slow but sure shift in the culture of our industry.

The building industry is large and will continue to accommodate a varied range of firm types and project delivery approaches. Both multidisciplinary integration and interdisciplinary collaboration will have a place, but scale and intent can be decisive: for large infrastructure projects, the former might work better, while for projects geared toward design excellence, it might be the latter. Of course there will be plenty of overlaps, and firms will continue to feature not one approach or the other, but something in between. For architects, the real opportunity is in claiming leadership of the collaborative/integrative process. As Scott Simpson of KlingStubbins states rather convincingly, "We're in the leadership business, with design as our medium!"

Notes

1. Thomas Fisher. "Can This Profession Be Saved?" *P/A* (February 1994).
2. John Gendall references Hanif Kara of AKT in "Engineering Success," *Architect* (January 1, 2009).
3. Gary Hamel and C.K. Prahalad. *Competing for the Future* (Boston: Harvard Business School Press, 1996).
4. American Institute of Architects—Code of Ethics and Professional Conduct 2007, http://www.aia.org/aiaucmp/groups/aia/documents/pdf/aiap074121.pdf, accessed May 2010.
5. Gendall. "Engineering Success."
6. Bernard Michael Boyle. "Architectural Practice in America, 1865–1965—Ideal and Reality," in *The Architect*, edited by Spiro Kostof (New York: Oxford University Press, 1977). Boyle references Walter Gropius's writing from *The Scope of Total Architecture*.
7. Ibid. Boyle references Nathaniel Owings's writing from *The Spaces in Between*.

8. Guy Nordenson referenced by Matt Chaban in "A New Era for an Old Partnership," *Architects' Newspaper* (December 11, 2006).
9. Excerpted from a class presentation for GSD 7440, "Leading the Design Firm," by Richard W. Jennings at the Harvard Graduate School of Design, September 23, 2009.
10. Andrew Pressman. "It's a Very Good Time to Develop Your Firm's Collaborative Skills," *Architectural Record* (April 2009).
11. Jay W. Lorsch and Thomas J. Tierney. *Aligning the Stars: How to Succeed When Professionals Drive Results* (Boston: Harvard Business School Press, 2002), p. 7.
12. Len Charney, head of practice at the Boston Architectural College, quoted in Pressman. "It's a Very Good Time to Develop Your Firm's Collaborative Skills."
13. Scott Simpson presented his views during a class discussion in GSD 7413, "Integrated Project Delivery," taught by Richard W. Jennings in Spring 2010 at the Harvard Graduate School of Design.
14. Hamel and Prahalad. "Competing for the Future."

Adelaide Wharf

Allford Hall Monaghan Morris / AKT

Collaboration as a Working Process

Morag Tait

Process is as vital to design as it is to production. When embarking on a project, all architects must define the rules of their engagement, and the limits and opportunities that exist. All pieces of work will then be a response that assimilates these constraints and opportunities.

—Alford Hall Monaghan Morris Project Manual, 2004

An Iterative Design Process

There is a common (mis)conception that an architectural design emerges as a pure and buildable concept, arriving in almost mystical fashion from the mind of a sole genius. This model is often supported in education, as individual students strive to produce the most evolved versions of their designs, to refine their proposals into the purest expressions of their concepts. It is often perpetuated by the image of the architect in popular culture, and some architects cheerfully reinforce this mystique. But even for the most brilliant of designers, design is a process, a series of steps that get to a workable solution without ever losing sight of the vision for the project. But how the process is shared, and what steps it breaks down into, varies from architect to architect.

Architecture can be considered as art allied to function: it fulfils a briefed purpose as well as being delightful. The architect's design skill comes in refining this building and program into an architecture with a coherent concept and execution. It requires a design vision, but one that has to resolve the functional aspects of a brief—spatially, structurally, financially, and in terms of construction. The build structure has to provide suitable spaces, allow people to use and inhabit it, be structurally sound, temper the environment—and be a good piece of architecture, offering delight to those who use it, visit it, and pass by it in the city.

Design is rarely a linear, predictable process but rather an iterative one, involving analysis and identification of the parameters set for the design, in terms of brief, site specifics, statutory requirements, budgets. Each iteration then responds to and tests ideas, refining them so that an appropriate design and vision emerges.

Collaboration

In architectural terms, collaboration involves the design disciplines working together, often with the architect as lead designer—sharing knowledge, learning from each other, and building a project that reflects a consensus that this is the best solution. Collaboration among disciplines is an approach that acknowledges the process of design development; it depends on the analysis of the problems and an iterative feedback of design solutions and options to the entire

The Model Scheme

The Site Specific Scheme at Adelaide Wharf

team, so that collective decisions are made at each stage. This requires a methodology of presentation where the logic of design moves is explained—transparent and open for comment.

In the case of Adelaide Wharf, collaboration involved the engagement of client and constructor, as well as the design team in a client-led research project on a theoretical urban site, where all parties bought into the process and result. This study, financed by the client, was subsequently used in a competitive bidding process that led to the development of the Adelaide Wharf site. The client set a brief to develop a system for low-cost one- and two-bedroom apartments. This involved presenting each design development for discussion, to be questioned, tested, and commented on—a valuable process, but not without its difficulties. Different working processes and expectations are exposed in a very direct way.

Collaboration with Client and Engineers

Designers generally understand the iterative nature of design, whether they are designing structures, services, or architecture. They expect that as information and parameters become clearer, potential solutions shift. This means that while each discipline is running its own process and feeding it back to the whole, it is not expected to be the final answer at each stage; a certain amount of abortive discussions are inevitable, and the process is not explicitly linear.

With regard to structural and services engineering, without early input on the constraints, architectural design development inevitably falls back on what architects already know, drawn from rules of thumb or precedent, to fill the gaps; or they design optimistically, without understanding the impact of design decisions, leading to compromises later in the process. Sometimes those rules of thumb hold true, when the starting point is similar to projects already done. In the case of Adelaide Wharf, the earlier research project (with the same collaborators) for the same client, First Base, had established these "rules" to a sufficient extent for the research model to provide a robust framework. The team had tested the design parameters, most of the brief was fairly clear for all parties, and the design was able to progress quickly. This meant that the team was working to established assumptions such as a grid size to suit the apartment module, a flat structural soffit, a repetition with an ease of manufacturing for the structure as well as the façade, and services zones established in the flat layouts for ease of installation and maintenance.

But each project is inevitably a new brief, and some of those model rules unraveled in the face of the messy reality of an actual site. This building had to turn corners, respond to an actual context; the apartment sizes and mix had to meet the criteria for the social landlord, the plan-

The diagrid structure at
the Yellow Building

Tooley Street internal spaces with
fair-faced finishes

ning authority, wheelchair users, and the private market as well as the potential inhabitants, all affecting assumed floor-plan sizes, layouts, and repetition. Inevitably some decisions had to be reconsidered, but the benefit was that with some principles established, the design team members were able to move forward on several fronts at once.

With the benefit of the research project behind the team, all were confident in working to those assumptions, knowing that the testing and discussions had already taken place. In most design development, without having examined the engineering alternatives and options available with the architectural design, opportunities are missed and the potential is not sufficiently pushed. Engineers tend to work on many more projects at a time than do architects, and so often are exposed to and have tested out their ideas on a broader range of projects and have examined new structural and construction techniques—their advantages and pitfalls. To be able to tap into this knowledge as the design evolves is invaluable.

Similarly, for engineers on Adelaide Wharf, seeing the architectural layouts evolving, hearing height discussions, allowed them to advise on effective grids, column types, and construction methodologies as the plans developed, so that the plans and sections were coordinated in principle and did not need reinventing later. Working closely from the start suggests that coordination between the disciplines can lead to better details, but it requires that all consultants update their information simultaneously as the design develops, all through the process. Without the input of the engineer, at the right times, the inevitable happens—downstand beams appear across flat soffits; vents, pipes, and grilles emerge willfully on carefully composed façades.

But perhaps the most vital aspect of collaboration is in creating buildings whose systems are so integrated they have to be designed together to exist at all. At the Yellow Building, the structural dia-grid wraps the edges of the slabs externally and internally around the atrium. The structure is used expressively as part of the architectural skin of the building, and the dynamic structural form dominates the reading of the space. Services, structure, and architecture are so integrated that they are enmeshed in each other in our project at Tooley Street. Using exposed concrete as an architectural finish, precast elements perform a structural function, but also distribute air through hollow columns, form the external skin, and are left exposed for thermal mass and a fair-faced ceiling finish.

Collaborating with the Contractor

In involving the contractor, there is potentially a wealth of knowledge to be drawn on including their understanding of construction techniques, both innovative and standard, to save time or money, and therefore improve quality within the budget. But the iterative world of early design development can be frustrating for them, as they would like to pin down the designer as soon as possible: each sketch is seen as the embryo that will evolve into the perfect solution. They are less comfortable with the process of sketches as evolutionary explorations, with forms changing radically at each step. They tend to want a tidy, linear process.

Of course, contractors do not usually have design backgrounds—they are organizers of processes, dealing in concrete, achievable elements. Their equivalent process is in breaking down the design into constructible packages, in procuring and programming, in how the parts get manufactured, and they find it difficult to engage in the shifting world of design development, to remain in the world of the speculative, for too long.

In the Adelaide Wharf process, this made them push for concurrent program stages, so that they could procure while the design team designed. This created a lot of tension, a lack of recognition of a necessary design process that frequently leads to dead ends, and abortive work for all parties. Equally, it made designers make firm decisions early on, which they then had to argue strongly about if they wanted to change them. The benefit was a perceived realism about time and cost estimates built into the process, enabling decisions to be made around the contractors' real cost base. In other schemes, such as volumetric ones such as Raines Dairy, the design can barely progress without their input, as the configuration of their system drives so many design decisions that the design process happens within their somewhat inflexible envelope. The bathroom pods at Adelaide Wharf were a similar situation.

Some trade contractors actively and positively engage with the process. At Adelaide Wharf, a specialist façade engineer oversaw an early tender to enable a choice of façade system, and subsequent cost negotiation was done on an open- book basis. The façade then evolved with close collaboration with the preferred façade manufacturer, as they developed their unitized system to suit our design as it evolved. They had to invent elements to suit our balconies, our window types, and most especially our rough-sawn timber cladding, a technology they had never applied to their system before. But they responded to our research with equivalent of their own and applied themselves to manufacturing the façade to the same levels of quality as their usual curtain-wall product.

Working with an engaged client and the contractor from the earliest stages meant that design team members had to be conscious about how they presented their logic. Starting by talking about options yet to be explored, the team acknowledged that decisions in each discipline affected others' work and sought buy-in for each step to come.

Expertise

Buildings are often now hugely complex constructions, financially (although this is not addressed in this text) as well as physically, and in terms of the statutory requirements surrounding them, with brief program descriptions. Each aspect of a brief requires its own expertise, and so in practice collaboration is ever-present in the process—with members of the design team, the client, and the contractor. Each collaborator is relied upon as the expert in his or her field.

Getting the most out of collaboration hinges on each party having this solid foundation as an expert, not a generalist. The more capable and skilled the participants, the better the collaboration is likely to be. The participants have to be good architects, engineers, or consultants, who empathize with and respect the other disciplines, to get the most out of each. In this manner, each pushes the others to improve the solutions offered, while pushing their own discipline to further the common design goal, to come up with ideas that might have gone unexplored had the collaboration not taken place.

Center for Advanced Architecture

Andrew Pedron, Nathan Shobe, Trisitie Tajima

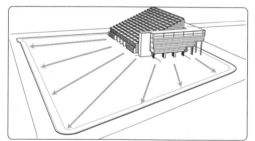

ENHANCEMENT OF EXISTING YARD CONDITION

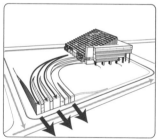

URBAN CONNECTIVITY

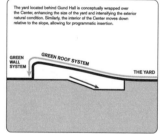

The yard located behind Gund Hall is conceptually wrapped over the Center, enhancing the size of the yard and intensifying the exterior natural condition. Similarly, the interior of the Center moves down relative to the slope, allowing for programmatic insertion.

GREEN WALL SYSTEM — GREEN ROOF SYSTEM — THE YARD

LANDSCAPE INUNDATION

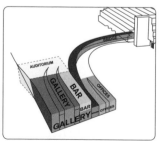

AUDITORIUM — GALLERY — BAR — OFFICES — GALLERY

PROGRAMMING

ANYWHERE

POTENCIAL TEMPORARY PAVILION

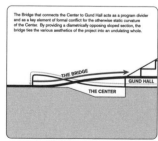

The Bridge that connects the Center to Gund Hall acts as a program divider and as a key element of formal conflict for the otherwise static curvature of the Center. By providing a diametrically opposing sloped section, the bridge ties the various aesthetics of the project into an undulating whole.

THE BRIDGE — GUND HALL — THE CENTER

BRIDGE INTEGRATION

Without interfering with the perceptual interface between Gund Hall and the surrounding context, while at the same time maintaining the openness of the yard, a building that houses a flexible structural and programmatic system is created. Although the given programmatic functions of the building required a large open gallery space, a lecture space, a bar, and service spaces, these spaces seemed rather at odds with one another. To accommodate this, sectional manipulation is used. By sinking the building underground in the back, it not only allowed for the grass of the yard to extend onto the roof of the building, but also enabled a certain sectional rake necessary for a functional lecture-hall space. Though this space is larger than what is necessary by programmatic needs, its primary strengths lie in its ability to accommodate gallery events through its large landings and flat walls.

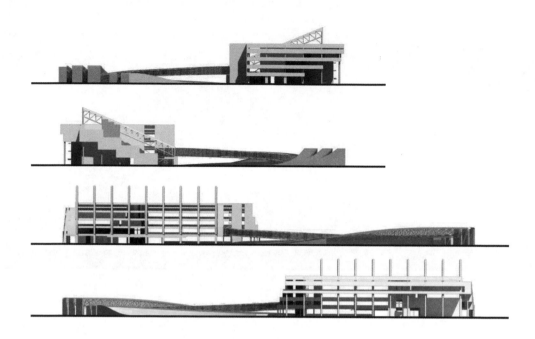

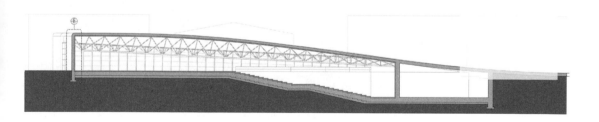

The gallery, an open space at the north of the building, is a light-controlled area that also operates as a lobby for the lecture hall and as a grand architectural space. The bar is suspended above all of this, allowing for direct interaction with the massive architectural edifice. The service functions, which include offices and conference rooms, are slightly separate from the primary public spaces, split by a sky bridge that connects the building to the third floor of Gund Hall. Through its shift upward, and the building's inset into the ground, a criss-crossing bifurcation that gives both formal complexity and structural complication is created.

INTERIOR LIGHTING STUDY

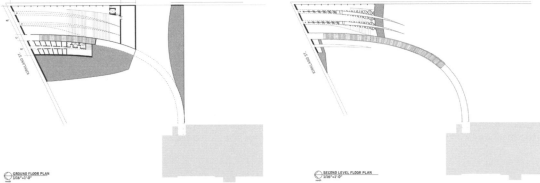

GROUND FLOOR PLAN
1/16"=1'-0"

SECOND LEVEL FLOOR PLAN
1/16"=1'-0"

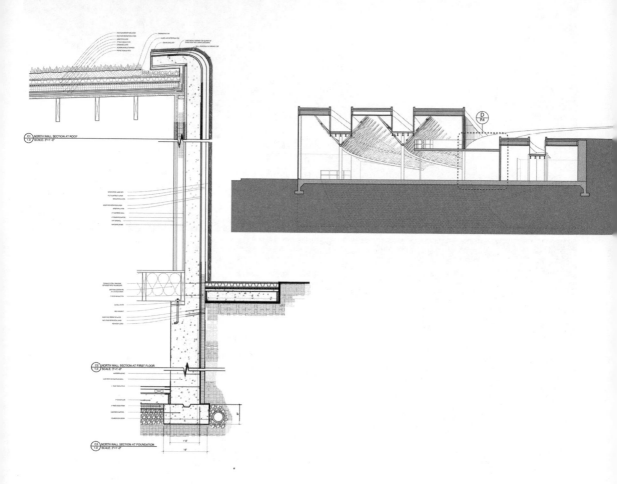

The structural system at play in the primary spaces is a one-way space-frame truss system that is essentially a massive triangle with the ground forming the hypotenuse. The structural system of the bridge can best be defined as a steel tube of weaved members that curves gently toward the system, creating a massive torque on the moment connections. This is alleviated by varying the degree and angle of the members as well as their density as they approach moments of greatest stress. Daylighting in the building is created by directing the glazing toward the east and west, providing light throughout the year, particularly in the winter when heat gain is minimal.

The green roof system facing the south allows for more controlled heat transfer and also helps to keep wind load to a minimum on the structure of the building. The green roof on the building, as occupiable space, also allows for continued use of the yard behind Gund Hall for temporary pavilions and outdoor exhibition spaces. By elevating these pavilions, the public can not only see them better but also can more directly engage in the green roof. The building thus becomes both a pedestal and a glass exhibition box for student work.

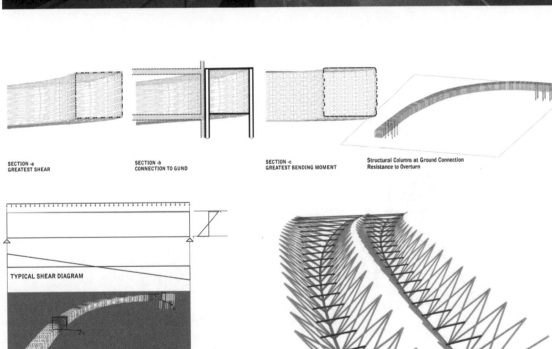

SECTION -a
GREATEST SHEAR

SECTION -b
CONNECTION TO GUND

SECTION -c
GREATEST BENDING MOMENT

Structural Columns at Ground Connection
Resistance to Overturn

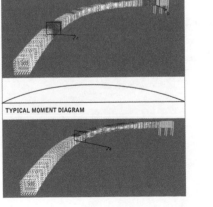

TYPICAL SHEAR DIAGRAM

TYPICAL MOMENT DIAGRAM

INTERIOR EXHIBITION HALL TRUSS SYSTEM

Bridge Gallery

Danxi Zou, Song He, and Bo Feng

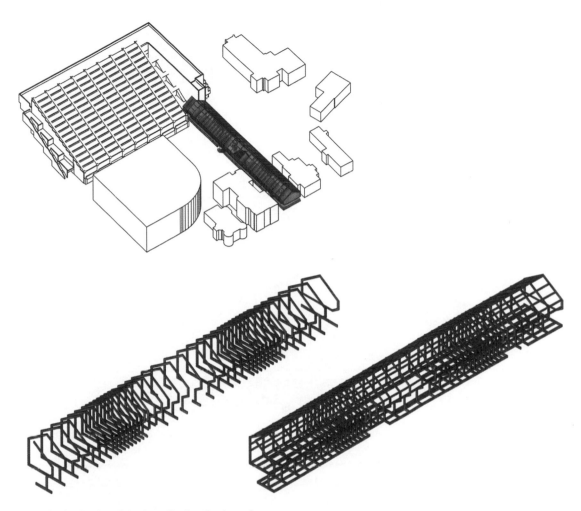

Change the density of wood structure rather than the shape of
the structure to satisfy mechanical requirements. The strength
of the structure comes from two tubes, the inner tube hangs
from the outer tube

The project stems from the idea of making a bridge the building itself. The linear volume,
floating above the ground and crossing the yard, makes a strong statement on the site
about its relationship with Gund Hall and the surrounding environment. Inside the building,
the tube serves as a main exhibition space for GSD students and faculty. A second tube
hanging from the roof serves as space for private events and administration. We chose
wood frame to construct both the outer and the inner tube—the floor, the façade, and the
roof. A pair of trusses supports the whole structure. By adjusting the density of the wood
frame according to load distribution calculations, the building achieves an optimal structural performance. Our concern for the project's sustainability is reflected in its minimal
interference with the existing building and landscape, its material (locally produced and
easily recycled) and its orientation in relation to the prevailing winter wind.

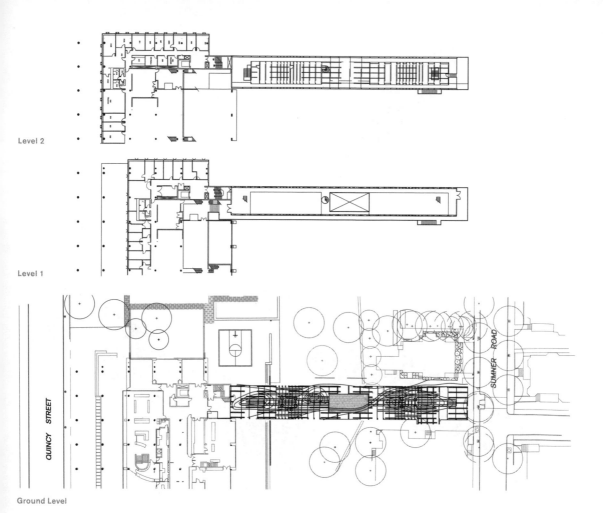

Level 2

Level 1

Ground Level

QUINCY STREET

SUMNER ROAD

Why a linear building?

The new building is much smaller than Gund Hall. How to make it fit into the context with a humble but attractive gesture became the main concern. Because a bridge connecting the two buildings is requested in the design guidelines, we decided to make the bridge the building itself, which is visually powerful from both exterior and interior. The linear volume, floating above the ground and crossing the yard, makes a strong statement about its relationship with Gund Hall and the surrounding environment.

Why at this location?

The site includes the backyard of Gund Hall and several private houses. Trying to maintain minimum interference with the existing buildings and landscape, we oriented the building in the east-west direction. The site offers a 75-meter-long corridor for the building to stretch out without touching any of the nearby houses. It would also keep the yard the same size, but this new structure above would provide new possibilities for using the yard.

Why a wooden frame?

At the beginning of this project, we considered other structure materials, such as concrete or steel. But we decided to make a wooden building for three reasons: first, compared with concrete or steel, wooden structures are more widely used in New England, especially in small-scale projects. Thus there are more local construction techniques that can be borrowed for our project. The construction quality is also easy to guarantee. Second, to make this building simple, wood is a good choice both for the structure and the façade. At the same time, it makes the building more harmonious with the surrounding wooden houses. Third, wood is a locally produced material. A wooden building can be easily demolished and recycled.

Structural density adjustment

To make the whole building more light, the supporting points were reduced to the minimal extent possible. The final result is two supporting systems, each holding a cantilever structure. To optimize the system structurally, more structure materials should be distributed near the supporting point. The usual approach is a structure with variable cross-sections. But in this project, we maintain equal cross-sections and achieve a similar structural performance by adjusting the density of the wood frame according to load distribution. More cross-sections and connecting material are arranged near the supports and less at the cantilever end. A second layer of structure following the same rules is supported from the main body to provide space for private events and administration.

Supporting pillar design

Each supporting system contains eight pillars, which connect the main structure with steel joints. Each pillar is made of four slightly curved wood pieces that bond in a tubular shape to get a better performance. The eight pillars are than fixed to a steel stand that slightly touches the landscape below.

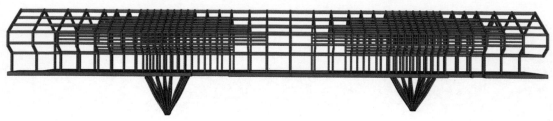

Density of structure distribution

Typical structure shape

Load distribution

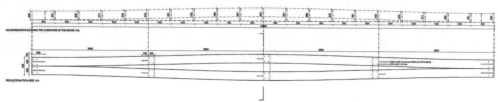

Supporting structure beam curvature

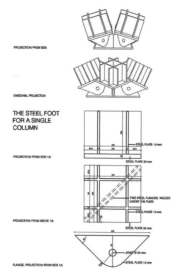

Detail reference

Supporting structure

Façade design

Five types of wooden panels with different window locations are used for the façade.
The window size is limited to maximize the exhibition space inside. The structure system is
partly exposed. The panel size follows the module of the main structure, so it can be easily
plugged in to the structural system. The panels are randomly sequenced to give the façade
an integrated but rich look.

Sustainability

The east-west orientation of the building could reduce the amount of the façade that faces
the prevailing winter wind from the west, decreasing the building's energy consumption.
The building uses skylights as its main lighting strategy. Exhibition space, requiring indirect
lighting is located at the low level and offices and public spaces are at the upper level.
The roof windows can be partly opened for natural ventilation.

View from the south between GSD and CGIS

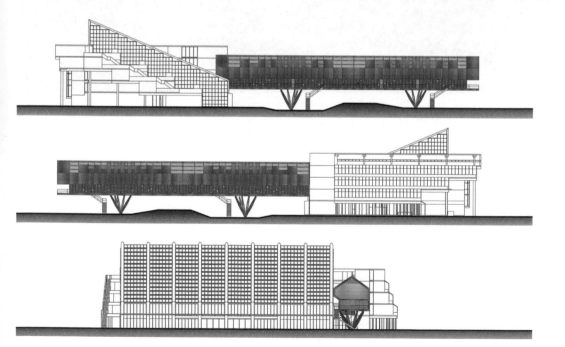

East end at Sumner Street

From Gund Hall Terrace

Westfield Student Village

Feilden Clegg Bradley Studios / AKT

Climate Comfort Collaboration

Ian Taylor

Achieving comfortable internal conditions alongside low-energy outcomes for buildings in a variety of climates depends on successful integration of the beliefs, knowledge, and processes of the design team. The initial approach to design will be underpinned by a belief in the necessity of an environmentally benign approach. There are many possible ways forward, which need to be discussed with the client. Maintaining appropriate internal comfort conditions with minimal carbon impact is a challenge that requires different solutions in different climatic conditions. These responses will tend toward the development of new vernacular forms—in contrast to a universal Modern movement aesthetic.

Believing in the argument, however, is of little merit without fundamental knowledge of the issues that affect the solution to these challenges. The architect must lead the design team toward a holistic design approach through knowledge of the relationships among climate, form, structure, and building services. Initial design decisions relating to building orientation, form, and passive environmental control can have a profound impact on the solutions. The methods of control and the level of active systems necessary to temper internal environments will be strongly influenced by the initial design decisions that affect passive performance.

An interactive design process is required to test concepts and strategies at an early stage and to inform the client and design team about the impact of early design decisions on the operation of the building. This process requires the design team to have greater levels of understanding of the interactions between separate disciplines' areas of expertise. The architect has to be more aware of the aesthetic impacts of exposed structure and services; the structural engineer has to be more aware of the thermal performance of the building frame, and of its embodied carbon impact; and the services engineer has to understand the role of the building fabric in moderating internal comfort conditions by passive means.

Four Stages of Interdisciplinary Design

1. Examining Preconceptions and Targeted Solutions

The initial stages of design are critical for establishing the direction of the project, and it is often useful to consider precedents while reviewing design parameters. Testing the brief against previously developed models and typologies can be a useful starting point to identify potentially fruitful strategies, or to dismiss blind alleys. The risk in this methodology is that a design solution is imposed on the project, without that solution having been generated through a fundamental analysis and prioritization of the brief and site opportunities. Such a preconceived design solution may address some of the site or functional issues, but contain inherent failings with regard to other issues particular to the new context. Targeting outcomes is a useful discipline at these early stages: describing the building's character and the terms of its operation can help prioritize the most important attributes of the design and crystallize thoughts on means to achieve these outcomes.

2. Prioritizing Issues

It is useful for the architect to employ tools to help the client and design team prioritize elements of the building's performance, to establish the focus of the design team effort. Within the context of bioclimatic issues, the design team will need to address diurnal and seasonal changes in the local environment of the site, investigating the impacts of temperature, wind, rain, sunlight, and acoustic conditions.

Feilden Clegg Bradley Studios have developed a matrix to describe the range of design targets and solutions available to the project to address particular environmental issues. By defining good-practice, best-practice, innovative, and pioneering benchmarks, this tool helps the client and design team prioritize those features that are felt to be the most significant for exemplary performance and innovation.

3. Directing the Design Process

Adherence to the design principles set out at the inception of the project is always a challenge in the face of practical, cost, and construction issues that arise during the process. The architect needs to lead the way through often conflicting demands. Our experience shows that significant design moves that integrate structure and services into the building fabric are more likely to be retained when all parties to the design process have been involved in their inception and the relationships between elements are understood. Some fundamental issues can become problematic when the design team is fractured—for example, in a design-build contract where a contractor takes over the lead of a project and brings in a mechanical services contractor before detailed design is complete.

4. Using Holistic Feedback

In the United Kingdom, there is increasing interest in the benefit of post-occupancy evaluation as a tool to monitor the performance of buildings in use and to provide data to optimize their operation. This feedback loop can be significant at early stages in the design process in reminding the design and client team to consider operational and control issues.

Toward a New Collaboration

The last eighty years have witnessed shifts in attitude toward building design and the control of the internal environment, reflecting the preoccupations of each decade. During this period a demonstrable lack of contextual climatic response developed alongside the pervasiveness of a debased Modern movement universal style. The environmental imperative that is now forcing an adjustment in how we use world resources is creating an opportunity to reassess the relationships among buildings, energy consumption, and lifestyle. A new holistic and integrated design process is required to address these challenges.

A somewhat simplistic review of the relationship between Western society and design over this period indicates the imperative to respond to changing attitudes to technology, offering insight into how this process should progress to provide better design now:

1930s A new aesthetic develops based on the promise of mass production and engineering in the face of economic gloom. This new aesthetic was influenced more by structural possibilities and ideas of social change than by environmental determinants.

1940s War.

1950s Recovery/new world optimism underpinned by technology: Time-saving household appliances for the home and technical solutions enable easier and cleaner heating and ventilation of buildings, facilitating a dislocation from traditional and vernacular forms by freeing architectural form from environmental constraints.

1960s Engineering and love can solve everything, in this era of moon landings and increased freedom. The prevalence of air conditioning and simple heating systems enable transfer of new building typologies, without amendment, into different environmental contexts across the world.

1970s Social awareness / people power: An alternative culture reacts against technological solutions developed alongside the mainstream.

1980s Money can solve everything in an individualized "me" culture. More sophisticated technological versions of the 1960s office fine tune the internal environment.

1990s Growing environmental consciousness. Exemplar projects investigate passive environmental control.

2000s Sustainability as necessity: Corporate and private acknowledgment of environmental issues pushes the agenda forward, ahead of the ability to deliver solutions.

2010s Sustainability is seen as positive goal to which design teams must respond.

Our current environmental dilemma requires an integrated design approach that responds to climate conditions by maximizing the passive control of the internal environment and using minimal resources in construction and operation. A new cultural climatic architecture should develop: to design within the means of our planet, responding to local environmental conditions, enhancing the sense of place, and respecting cultural tradition.

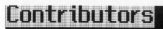

Contributors

Hanif Kara is a practicing structural engineer, the Pierce Anderson Lecturer in Creative Engineering at Harvard University's Graduate School of Design, and Visiting Professor for Architectural Technology at the KTH Royal Institute of Technology in Stockholm, Sweden, since 2008. As design director and cofounder of Adams Kara Taylor in 1996, his curiosity and "design-led" approach with interests in innovative form, material uses, and complex analysis methods has allowed him to work on award-winning, pioneering projects. His career extends beyond the structural engineering disciplines and led to his appointment in 2008 as a member of the Commission for Architecture and the Built Environment—the first engineer to hold this post for the government watchdog that monitors the quality of design throughout the United Kingdom. In March 2007 he was appointed as one of fifteen members of the Design for London Advisory Group to the Mayor of London. His constructed work is recognized as being linked with the research and education areas of design. He has been teaching at architectural schools in Britain, Europe, and the United States since 1996. He co-tutored a diploma unit at the Architectural Association in London from 2000 to 2004, acted as a consultant for the AA Design Research Lab for a number of years, and was an external examiner at the AA from 2005 to 2008. Hanif was selected for the master jury for the 2004 cycle of the Aga Khan Award for Architecture and served as a project reviewer in 2007 and 2010. He was awarded an honorary fellowship of the Royal Institute of British Architects in 2007. Hanif is also on the board of trustees of the Architecture Foundation, and in 2011 he received the ACE Engineering Ambassador of the Year Award. He was on the jury for the RIBA Stirling Prize 2011.

Andreas Georgoulias is a lecturer in architecture and research director for the Zofnass Program for Sustainable Infrastructure at the Harvard Graduate School of Design. He has worked in design and construction management with Obermeyer, Hochtief, and the U.S. General Services Administration, and in infrastructure financing with UniCredit. He has consulted for the Economist Intelligence Unit and he has been a project member for new city developments in the eastern province of Saudi Arabia and in Karachi, Pakistan. Recently he has been a consultant for the Greek government on the privatization of a 6,500-acre real estate asset in Athens. Andreas's research focuses on the interplay between project-level dynamics and people, technology, and systems of organization; he combines theories and methods from design, planning, organizational science, strategy, and psychology to explain performance variation in building projects, design firms, and urban developments. His current research includes the development of the Gulf Encyclopedia for Sustainable Urbanism, sponsored by the Qatar Foundation. His work on sustainable infrastructure assessment has been supported by an AEC industry consortium and conducted in collaboration with several professional societies, such as the American Society of Civil Engineers and the American Public Works Association. On this research he collaborates with the U.S. Environmental Protection Agency, the New York Department of Transportation, and the New York/New Jersey Port Authority. As part of the Zofnass Program he has presented his findings at the White House. His recent books include the co-edited *Infrastructure Sustainability and Design* (Routledge). He is the author of more than a dozen case studies used for teaching at Harvard, and he has presented his research at numerous academic conferences and professional venues in countries around the world. Andreas holds a degree in architecture engineering from the National Technical University of Athens, and a master's in design studies and a doctorate of design from the GSD.

Fai Au is the founder and principal of O Studio Architects. He received his master's in design studies from the Harvard University Graduate School of Design and his bachelor's of architecture (Hons.) from the Royal Melbourne Institute of Technology. He is a registered architect in Hong Kong and a member of the Hong Kong Institute of Architects. He is the winner of the 2007 Hong Kong Young Architects Award and was nominated for the 2010 China Architecture Media Award, Young Architect Prize. Fai Au previously practiced as senior designer at the Office for Metropolitan Architecture in Rotterdam. He was also the founding director of ADARC Associates Limited from 2005 to 2008. In 2009 he founded O Studio Architects and led his design team to deliver various award-winning projects including: the Church of Seed, the San Diego East Village Art District Development, and the Fanling Classic Car Museum. He previously taught master of architecture design studio at the Chinese University of Hong Kong and served as guest critic for design reviews there and at the University of Hong Kong. He is currently pursuing a master's of art in philosophy at the Chinese University of Hong Kong.

Jennifer Bonner is the director of Studio BONNER in Los Angeles. She received a bachelor of architecture degree from Auburn University and a master's of architecture from Harvard University's Graduate School of Design, where she was awarded the James Templeton Kelley Prize. She has taught design studios and seminars at Woodbury University, Georgia Institute of Technology, Auburn University, Lund University, and the Architectural Association. Her work has received numerous awards, including an AR Award for Emerging Architecture in 2005, and has been exhibited at institutions in Hollywood, Cincinnati, London, and Barcelona.

Since joining Heatherwick Studio in 2007, **Katerina Dionysopoulou** was the project architect for the UK Pavilion for the 2010 Shanghai World Expo. More than 5 million people visited the Seed Cathedral, and it was awarded the prestigious Lubetkin Prize by the RIBA. She recently led a team on the design of two parks in the Middle East, both significant public spaces designed for the challenging local climate. Katerina completed her diploma in architecture from the Aristotelian University of Thessaloniki in 2001. A year later, she completed a master's degree in architectural design with Peter Cook at the Bartlett School of Architecture, London. In the last four years, Katerina has taught at the Architectural Association and given numerous lectures worldwide.

Sylvia Feng is a designer currently working in New York at an international architecture firm specializing in the design and research development of education, lifestyle, and workplace spaces and structures. She has also worked as an exhibits and furniture designer. No matter the scale of operation, her aspirations and interest lie in design that is efficient in material, environmentally intelligent, and intuitive to users. She believes that design is primarily an exercise to solve problems in an innovative fashion, whether in architecture, engineering, or communications. Sylvia received her bachelor's of science in architecture (summa cum laude) from the State University of New York at Buffalo. She attended the Harvard Graduate School of Design, where she earned a master's in architecture I in 2009.

Lee-Su Huang received his bachelor of architecture degree from Feng-Chia University in Taiwan and his master in architecture degree from Harvard University's Graduate School of Design. He has practiced in Taiwan and the United States with Preston Scott Cohen Inc. in Cambridge, and on location with LA.S.S.A Architects in Seoul, Korea. Lee-Su is currently assistant professor at the School of Architecture, College of Design, Construction, and Planning at the University of Florida. He has also taught at the Boston Architectural College. His current research centers on digital design and fabrication methodology, genetic algorithms/optimization for architectural and urban parametric applications, and theorizing/conceptualizing of parametric design and systemic processes in architectural education as well as practice. He is also cofounder of SHo, a speculative practice that engages the crossover territory of digital design and fabrication and material experimentation, while exploring the broader implications of digital methodologies in architectural design.

Evangelos Kotsioris is a PhD candidate in the history and theory program at the Princeton School of Architecture. His research interests involve the intersections of postwar modernist architecture of the Soviet Union and information technologies. He graduated with first-class honors from the School of Architecture of the Aristotle University of Thessaloniki (2009), studied as an exchange student at the faculty of architecture of TU Delft (2007), and earned his post-professional master's in architecture II from the Harvard Graduate School of Design (2011). He has worked as an architect in Greece (2003–09) and Holland at OMA (2007–08). In parallel to practice, he worked as a research assistant at AUTh and Harvard GSD. He has taught introductory courses in the history and theory of architecture and architectural photography at the GSD (2010–11), as well as led design studios at the GSD and the Boston Architectural College (2011). Evangelos's work has been featured in international exhibitions such as the Venice Biennale of Architecture (2006) and in magazines such as Mark, Frame, and Conditions. His essays and articles have been presented at architectural conferences at Harvard and Columbia universities as well as in specialized publications.

George L. Legendre is a founding partner of IJP, a London-based practice exploring the natural intersection between space, mathematics, and computation. He graduated from the Harvard Graduate School of Design in 1994 and served as lecturer and assistant professor of architecture there from 1995 to 2000. Prior to founding his practice, he was visiting professor at ETH Zurich (2000) and Princeton University (2003–05), and unit master of diploma unit 5 of the AA School of Architecture in London (2002–08). IJP won a competition (with Adams Kara Taylor) to cover a central London street with 1,000 square meters of glass and completed (with RSP) Henderson Waves, a 1,000-foot-long bridge in Singapore. IJP served as delivery architect for the Bat House, a high-tech, sustainable shelter recently completed in the London Wetlands Centre. Among George's publications are *Pasta by Design* (Thames and Hudson, 2011), *IJP: The Book of Surfaces*, *Bodyline: The End of our Meta-Mechanical Body*, and the main critical essay in *Mathematical Form: John Pickering and the Architecture of the Inversion Principle* (all AA Publications, 2003–06). He guest-edited a special issue of *AD* magazine on the mathematics of space.

Sabrina M. Leon received a BSE in systems engineering from the University of Pennsylvania, with minors in architecture and mathematics. She was vice president of her school's American Society of Civil Engineers chapter and secretary of the Society of Hispanic Professional Engineers, and she worked as an independent design consultant for Wharton's Innovation and Talent Seekers. She was awarded the Engineering Faculty Appreciation Award at graduation. Sabrina's professional career took a turn toward architecture soon after college graduation. Though she interned as an engineer for CSA Northamerica in Philadelphia and Techint in Buenos Aires, she had always considered developing her design skills. She worked at NXG Architecture, a small boutique residential/commercial architecture firm in Florida, and quickly excelled as an interdisciplinary designer. She later went on to receive her master's in architecture at the Harvard University Graduate School of Design, where she continued to explore design as an amalgamation of engineering and architecture.

Murat Mutlu is the head of the architecture and design practice INOA|International Office of Architects, based in Cambridge, Massachusetts. His firm has had a number of commissions in the northeast of the United States and overseas. Previously he worked at Zaha Hadid Architects in London and at Skidmore Owings & Merrill in New York.
Murat is a graduate of the Cornell University College of Architecture, Art, and Planning, where he earned his bachelor of architecture degree. He continued his education at Massachusetts Institute of Technology, where he received a master of science degree from the School of Architecture and Planning. His research at MIT focused on computation and design.

Guy Nevill's engineering focus has been on passive design principles and how people occupy their buildings—thinking about the users' experience of a space in terms of both comfort and simplicity of use. He has been involved in a series of green HQ office projects including the HQ's for the National Trust and the Woodland Trust, with information gathered from post-occupancy studies used to test innovations and inform later designs. He has also been working on ways of improving the effective communication of sustainability among design teams and clients. He helped to publish the sustainability matrix tool in the *Architects Journal* to encourage this discussion, and he also contributed to an online design guide for architects. Guy is a senior partner at Max Fordham and also belongs to the British Council for Offices Environmental Sustainability Group.

Azadeh Omidfar is a master of design studies graduate from Harvard University's Graduate School of Design, where she was the 2011 recipient of the Daniel L. Schodek Award for Technology and Sustainability. Her research focused on building skins' performance and fabrication. She earned a bachelor's of architecture (with distinction) from California College of the Arts in San Francisco in 2008, where she received the technology book award. Azadeh has assisted in numerous design studios and technology courses at the GSD and California College of the Arts. She has been involved in researching sustainable infrastructure and with her

colleagues has created the initial sustainable rating system for national and international infrastructure projects. Azadeh has been a LEED-accredited professional since 2008.

Dimitris Papanikolaou, an inventor, designer, and maker, is currently a doctor of design student at the Harvard Graduate School of Design. He has been a researcher at the Smart Cities group at MIT's Media Lab, where he will earn a master's in media arts and sciences this summer. Dimitris also received a master's in design and computation from MIT as a Fulbright scholar, and a diploma in architectural engineering from the National Technical University of Athens in Greece, where he worked professionally as a licensed architect. Dimitri's multidisciplinary work integrates fabrication, digital/physical computing, and behavioral economics to create responsive infrastructure ecosystems. At the Media Lab he co-developed Mobility on Demand, a self-organizing vehicle-sharing system that uses incentive mechanisms to optimize performance. Dimitris has taught courses and organized workshops, exhibitions, and lectures across the United States and Europe. His work has been part of peer-reviewed conferences such as the International Conference on Complex Systems, MIT's Computational Sustainability, and the eCAADe Conference; exhibitions such as "Ecological Urbanism" at the GSD and the Icsid World Design Congress in Singapore; and books such as *Reinventing the Automobile* (MIT Press) and Infrastructure *Sustainability and Design* (Routledge).

Christos Passas is an associate director at Zaha Hadid Architects. He has worked in the firm as a senior designer since February 1998. He was project architect for numerous projects and winning competitions, including most notably the Stirling Prize nominee, the Phaeno Science Center Wolfsburg, completed in November 2005. Christos gained his professional degree, BArch (Hons.), at the School of Architecture of Pratt Institute in 1995, as a Fulbright scholar. He continued his postgraduate studies in the field of advanced architectural design at the Architectural Association, London, where he received his graduate design diploma in 1998. Since 2007, he has returned to academia and is now involved in teaching parametric design and urbanism at the Architectural Association and is Guest DAAD Professor in the Dessau Institute of Architecture at Bauhaus (Dessau), where he holds his own master design studio.

Burak Pekoglu, after graduating from Robert College in Istanbul, received a bachelor of science degree in architecture from the State University of New York at Buffalo, where he attended an exchange program at the Aarhus School of Architecture in Denmark, and a

Master of architecture degree from the Harvard Graduate School of Design in 2011. His earlier art-related involvements include an apprenticeship with sculptor Irfan Korkmazlar and a solo sculpture exhibition and collaboration with Ronald Lopez on an artist exchange program and various art festival organizations in Istanbul. He worked at SOM's New York office, Schmidt Hammer Lessen's Aarhus Office, and Mimarlar Tasarım in Istanbul as an intern architect. In 2009 he worked at TACA's Harbiye Istanbul Convention Center construction and completed the design-build of a 240- square-meter contemporary art gallery. At the GSD, he was teaching assistant to Hashim Sarkis on two studios sited in Istanbul and to Edwin Chan on the studio "Convergence: The Museum Artistically Reconsidered," sited in Los Angeles. His thesis covered Eskisehir's train station project, analyzing infrastructure's impact in the hinterland of Turkey. He founded BINAA (Building INnovation Arts Architecture) in 2012 as a collaborative platform where innovative and artistic ideas find form. He currently works at Pelli Clarke Pelli Architects in New Haven, Connecticut.

Fernando Pereira Mosqueira is a civil engineer who works as part of the Optimization Technical Center at AIRBUS. After completing his degree on civil engineering (Ingeniero de Caminos, Canales y Puertos) at University of La Coruna, he pursued a master's of engineering in high-performance structures at MIT. His work as an engineer includes fluid–structure interaction, structural dynamics, and aircraft structures optimization.

Lana Potapova is a bridge engineer in Arup's New York bridge group. She joined Arup in 2009 after earning her master of engineering degree from the Massachusetts Institute of Technology. While her thesis work revolved around bridge construction, she also developed a strong interest in efficient delivery mechanisms of structures with complex geometries. Her supplementary coursework at the Harvard Graduate School of Design and MIT School of Architecture exposed her to innovative tools and tactics that facilitate the communication between engineers and architects, and expedite the design process of atypical structures. Lana's expertise in parametric modeling and computational geometry tools allow for the efficient design of large infrastructure projects and facilitate the delivery of bridges with complex geometries. Lana now leads the skills development area for Arup's New York bridge group.

Advait Sambhare is an architect and urban designer based in Mumbai, where he is associated with G. D. Sambhare & Co. (GDS), an architectural practice catering to diverse projects in the private as well as public

sector. In addition to leading studio teams, he is also responsible for facilitating GDS's evolving strategic collaborations with international design practices, outreach programs to various schools of architecture, in-house skills development programs, and project- and practice-related management reviews. Advait graduated from Harvard University's Graduate School of Design in 2010, where he pursued his interest in design leadership and interdisciplinary collaboration in the master in design studies program in project management and real estate. Advait previously obtained a master's degree in urban design from the University of Michigan and a bachelor's degree in architecture from Sir J. J. College of Architecture, Mumbai, where he has also served as a visiting studio instructor. He has been invited to various architecture schools as an external critic for design reviews. Advait is registered with the Council of Architecture in India.

Scott Silverstein is a structural engineer at The Louis Berger Group, Inc., in Needham, Massachusetts. He holds a bachelor of science degree from Cornell University and a master's of engineering in civil and environmental engineering from MIT, with a focus on high-performance structures. Scott has performed numerous inspections and condition evaluations of deteriorated structures such as building façades, roofs, and bridges. He has also worked on new structural designs that incorporate principles of adaptability, including the North Bank Bridge, a landmark pedestrian bridge in Boston that reuses the foundations of a demolished highway on-ramp. Scott's research interests include life-cycle design and nontraditional building materials.

Jessica Sundberg Zofchak currently works as an environmental designer in New York City, where she focuses on daylight design and façade optimization through quantitative and computational analysis, while consulting on general sustainability initiatives for a variety of project types. She seeks to produce high-quality structures that challenge conventional approaches by using integrated, innovative, and pragmatic design. Her background includes bachelor's of science degrees from the Massachusetts Institute of Technology in building technology and marketing science, as well as a master's in high-performance structures.

Morag Tait joined Allford Hall Monaghan Morris in 1999. She has worked on a wide variety of housing, commercial, and health-care projects, including the first housing scheme for public-sector workers to be realized under the Government's London-Wide Initiative at Adelaide Wharf, Hackney. She is currently directing projects across a range of sectors, including Norwood Hall Joint Services Centre, a health and leisure center, as well as mixed-use housing projects in Camden and

Stoke Newington. Morag was made an associate director in 2005. She coordinates and chairs the practice's regular project architect reporting meetings and has formal responsibility for internal design stage reviews.

Ian Taylor joined Feilden Clegg Bradley in July 2001 and became a partner in January 2003. He previously worked at Arup Associates before becoming an associate director at Bennetts Associates, where his work on the Wessex Water Operations Centre, the Richard Attenborough Centre at Leicester University, Jubilee Library Brighton, and proposals for an extended Museum of the Moving Image for the British Film Institute in London set new standards for environmental performance and inclusive design. Ian has been involved with the Building Research Establishment on the development of BREEAM and Envest environment analysis tools. He has worked on eleven RIBA-award-winning projects, two of which reached the Stirling Prize shortlist. He has recently completed student housing for Queen Mary University of London, the Northampton and Islington Academies, and the One Planet Living Community in Brighton. He is currently working on projects at the London School of Economics, the University of Southampton, and the University of Washington in Seattle. He leads the FCBS Research and Innovation Group.

Chris Wan is the manager of city design at Masdar City, a major sustainable development in Abu Dhabi. Chris is a registered architect in the United Kingdom and Hong Kong. As an architect, he has worked for Richard Rogers Partnership and David Chipperfield and Partners in London and for Rocco Design Ltd. in Hong Kong. Later he moved to the client side of the building industry by working for Walt Disney Imagineering Hong Kong and Sorouh Real Estate in Abu Dhabi before joining Masdar City. Chris has given lectures on the subject of sustainable architecture at the Masdar Institute, the Paris-Sorbonne University Abu Dhabi, and the Institute for Advanced Architecture of Catalonia. He holds BSc and BArch degrees from the University of Bath.

Dan Weissman is a master of design studies candidate at Harvard's Graduate School of Design and a studio instructor at the Boston Architectural College. Dan's research focuses on the relationship between digital tools and human perception, and their integration into architectural pedagogy, as well as the role of passive environmental strategies in emerging and informal urban conditions through a triple-bottom-line sustainable development in post-earthquake Port-au-Prince, Haiti. Dan was the 2010 recipient of the AIA Henry Adams Medal at the University of Michigan, where he received a master of architecture degree and was integrally involved in the university's campus sustainability initiative. Before entering graduate studies, Dan worked as an architectural lighting designer at Lam Partners in Cambridge, Massachusetts, where he employed digital analysis for many projects, learning first-hand the opportunities as well as the limitations of digital tools as an integral element of the design process. He has taught architecture and lighting design in numerous settings including the Boston Architectural College, the University of Michigan, and the GSD.

Acknowledgements

Interdisciplinary design is as much about people as it is about design, processes, and buildings. It is the human intellect, spirit, and soul that create the magic in our field. Likewise in this publication: the unique qualities of all of the people who met, interacted, and worked together, eventually created the beautiful things you see in this book. We would like to express our deep gratitude to Mohsen Mostafavi, Dean of the Harvard University Graduate School of Design, Preston Scott Cohen, Chair of the Department of Architecture, and Patricia Roberts, Executive Dean of the GSD, for making this project a reality; Spiro Pollalis, Professor of Design, Technology, and Management at GSD, for starting the course with us in 2007; Melissa Vaughn, GSD Publications Director, for her dedicated work and support throughout the publication process; Albert Ferré and Ulises Chamorro from Actar for their excellent work on this book; Anthony Sullivan, Evangelos Kotsioris, Libby Farley, and Mark Watabe for working with us in developing this publication; Jeremy Melvin for providing valuable feedback on Hanif Kara's essay; Adams Kara Taylor overall for their support in many different ways; and finally all of the students and guest speakers of GSD6328: "In Search of Design through Engineers," for making such a great course. To all of you, thank you.

Image credits

3. Learning from Design, Andreas Georgoulias
Figures , 2, 3. I. Chistiakov, I. Lopes, M. Mutlu, S. Silverstein, H. Young (2009)
Figures 4, 5, 6, 7, 8. R. Fang, L. Aldana, S. Silverstein, M. Mutlu, M. Watabe (2009)
Figures 9, 10, 11, 12, 13, 14, 15, 16. S. He, C. Pedron, Y. Wang, A. Sullivan, J. Morgan (2011)
Figures 17, 18, 19, 20. H. Wu, L. McTague, Z. Wu (2011)
Figures 21, 22, 23. J. Bonner, M. Huang, L.S. Huang, J. Minguez (2008)
Figures 24. F. Au, S. Leon, A. McGee, F. Mosqueira, M. Ruettinger (2010)
Figures 25a, 25b. E. Kotsioris, A. Taylor, B. Pekoglu, K. Won (2010)
Figures 25c. J. Han, M. Holmquist, I. Lopes, A. Sambhare, H. Young (2009)
Figures 25d. F. Au, S. Leon, A. McGee, F. Mosqueira, M. Ruettinger (2010)
Figures 26, 27a, 27b, 27c, 28, 29a, 29b. T. Bost, S. Chang, J.S. Park, A. Pedron, D.Zou (2011)
Figures 30a, 30b, 31,32, 33, 34, 35, 36. V. Baranova, S. Morrison, N. Shobe, S. He (2011)
Figures 37a – f. D. Papanikolaou (2008)
Figures 38, 39, 40, 41, 42, 43, 44. R. Idris, M. Imbern, M. Scarlett, P. Semmler (2012)
Figures 45, 45, 46, 47. L. Fan, M. Gomez, R. Hawton, M. Koch, A. Puras, B. Shin (2012)
Figure 48. I. Ignatakis (2008)
Figures 49, 50. M. Mutlu, L. Potapova, S. Silverstein, J. Sundberg, M. Watabe (2009)
Figures 51, 52,53a, 53b, 53c, 53d. M. Watabe (2010)

4. Structure as Architectural Intervention, Sabrina Leon
Figure 1. Photo used under Creative Commons from Bruno Girin
Figure 2. Photo used under Creative Commons from jolyon_russ
Figure 3. Photo used under Creative Commons from greenmarlin
Figure 4. Photo used under Creative Commons from ihavegotthestyle
Figure 5. Photo used under Creative Commons from darkensiva
Figure 6. Photo used under Creative Commons from paenguin
Figure 7. Photo used under Creative Commons from stephenhanafin
Figure 8. Photo used under Creative Commons from Richard Moross
Figure 9. Photo by Gianni Berengo Gardin, courtesy of RPBW
Figure 10. Photo by Gianni Berengo Gardin, courtesy of RPBW
Figure 11.. Photo used under Creative Commons from karen horton
Figure 12. Photo courtesy of Richard Horden
Figure 13. Photo by Dennis Gilbert, courtesy of Richard Horden
Figure 14. Photo used under Creative Commons from End User
Figure 15. Photo used under Creative Commons from jeanbaptiste maurice
Figure 16. Photo used under Creative Commons from harshlight
Figure 17. Photo used under Creative Commons from areta
Figure 18. Photo used under Creative Commons from sanjoyg
Figure 19. Photo used under Creative Commons from kindee
Figure 20. used under Creative Commons from Brian Robert Marshall
Figure 21. Photo used under Creative Commons from zimpenfish
Figure 22. Photo used under Creative Commons from Simon Blackley
Figure 23. Photo used under Creative Commons from Simon Blackley
Figure 24. Image by author, after A. Ogg, (1987). *Architecture and Steel: The Australian Context*, p. 49.
Figure 25. Photo courtesy of Jourda Architectes
Figure 26. Photo used under Creative Commons from 14646075@N03
Figure 27. Photo used under Creative Commons from 14646075@N03
Figure 28. Photo used under Creative Commons from 14646075@N03
Figure 29. Photo used under Creative Commons from JoeInSouthernCA
Figure 30. Photo used under Creative Commons from Zol87
Figure 31. Photo used under Creative Commons from axzm1
Figure 32a. Photo used under Creative Commons from martinrp
Figure 32b. Photo used under Creative Commons from Oxyman
Figure 33. Photo used under Creative Commons from vito7
Figure 34. Photo used under Creative Commons from mueritz
Figure 35. Photo used under Creative Commons from endworld
Figure 36. Photo used under Creative Commons from lostseouls
Figure 37. Photo used under Creative Commons from tvol
Figure 38. Photo used under Creative Commons from Payton Chung
Figure 39. Photo used under Creative Commons from glokbell

Figure 40. Photo used under Creative Commons from IM Thayer
Figure 41. Photo by Bitterbredt, courtesy of Studio Daniel Libeskind
Figure 42. Photo used under Creative Commons from DVD R W
Figure 43. Photo used under Creative Commons from generated
Figure 44. Photo used under Creative Commons from Rodge500
Figure 45. Photo used under Creative Commons from art_es_anna
Figure 46. Photo used under Creative Commons from local_louisville
Figure 47. Photo used under Creative Commons from Voyager
Figure 48. Photo used under Creative Commons from Christian Wendling
Figure 49. Photo used under Creative Commons from Johann H. Addicks
Figure 50. Photo used under Creative Commons from Mbdortmund
Figure 51. Photo used under Creative Commons from glasseyes view

6. Surfing the Wave, Murat Mutlu
Figure 1. Image courtesy of ARUP
Figure 2. Image by author.
Figure 3. Image courtesy of AKT
Figure 4. Image by author

7. Toward a New Sobriety: Rebel Engineering with a Cause, Evangelos Kotsioris
Figure 1. Image courtesy of Arup
Figure 2. Photo used under Creative Commons from atelier_flir)
Figure 3. Image courtesy of ARUP
Figure 4. Image courtesy of REX
Figure 5. Image courtesy of OMA.
Figure 6. Photo used under Creative Commons from bfishadow
Figure 7. Image courtesy of Ali Rahim and Hina Jamelle/Contemporary Architecture Practice. New York.
Figure 8. Courtesy of Zaha Hadid Architects.

8. Highcross Department Store, FOA/ AKT
Figure 1. Image copyright Trimedia, courtesy of AKT
Figure 2. Image copyright Trimedia, courtesy of AKT
Figure 3. Image copyright Satoru Mishima, courtesy of AKT
Figure 4. Image copyright Satoru Mishima, courtesy of AKT

9. Ravensbourne College of Art and Design, FOA/ AKT
Figure 1. Image copyright Valerie Bennett, courtesy of AKT
Figure 2. Image copyright Valerie Bennett, courtesy of AKT
Figure 3. Image copyright Valerie Bennett, courtesy of AKT

10. GSD Lite, Francisco Izquierdo, Giorgi Khmaladze, Jarrad Morgan, Stephanie Morrison
All images by authors

11. Harvard MODE, Travis Bost and Werner Van Vuuren
All images by authors

12. Phaeno Science Center, Zaha Hadid Architects/ AKT
Figure 1. Image copyright Klemens Ortmeyer, courtesy of AKT
Figure 2. Image copyright Klemens Ortmeyer, courtesy of AKT
Figure 3. Image copyright Klemens Ortmeyer, courtesy of AKT

13. Design the Cloud: Cross-Disciplinary Approach to Design, Christos Passas
All images courtesy of Christos Passas/ Zaha Hadid Architects.

14. Changing Forms, Changing Processes, Dimitris Papanikolaou
Figure 1. Photo used under Creative Commons from Jaume d'Urgell
Figure 2. Photo used under Creative Commons from jimmiehomeschoolmom
Figure 3. Image courtesy of Lary Saas
Figure 4. Photo used under Creative Commons from T. Carrigan
All other images by author

15. Enhancing Prefabrication, Sylvia Feng
Image 1. Photo used under Creative Commons from shauni
Image 2. Photo used under Creative Commons from Energetic Spirit
Image 3. Photo used under Creative Commons from Taxiarchos228

Image 4. Photo used under Creative Commons from yusunkwon
Image 5. Photo used under Creative Commons from Chris 73
Image 6. Photo courtesy of Behrokh Khoshnevis
Image 7. Image courtesy of Behrokh Khoshnevis

16. Design for Disassembly: Closing the Materials Loop without Sacrificing Form, Scott Silverstein
Image 1. Photo used under Creative Commons from marcteer
Image 2. Image by author
Image 3. Image by author
Image 4. Photo used under Creative Commons from D'Arcy Norman

17. Design with Climate: The Role of Digital Tools in Computational Analysis of Site-Specific Architecture, Azadeh Omidfar and Dan Weissman
All images courtesy of Holly Wasilowski and Christoph Reinhart

18. Masdar Institute of Science and Technology, Foster + Partners/ AKT
Figure 1. Image copyright Peter Hind, courtesy of AKT
Figure 2. Image copyright Peter Hind, courtesy of AKT
Figure 3. Image copyright Peter Hind, courtesy of AKT
Figure 4. Image copyright Nigel Young / Foster + Partners, courtesy of AKT
Figure 5. Image copyright Nigel Young / Foster + Partners, courtesy of AKT
Figure 6. Image copyright Nigel Young / Foster + Partners, courtesy of AKT

20. Open House, Cara Liberatore, Leslie McTague, Anthony Sullivan
All images by authors

21. Refabrication, Charles Harris and Ji Seok Park
All images by authors

22. Heelis National Trust Headquarters, FCB Studios/ AKT
Figure 1. Image copyright Heliview Ltd, courtesy of AKT
Figure 2. Image copyright Dennis Gilbert / VIEW, courtesy of AKT

23. On Heelis: The Role Played by Environmental Engineers, Guy Nevill
Image 1. Image courtesy of FCB Studios/ Max Fordham
Image 2. Image courtesy of FCB Studios/ Max Fordham
Image 3. Image courtesy of FCB Studios/ Max Fordham
Image 4. Image courtesy of FCB Studios/ Max Fordham
Image 5. Image courtesy of Max Fordham

24. Algorithms in Design: Uses, Limitations, and Development, Lee-Su Huang
Figure 1. Photo used under Creative Commons from Jorge Lascar
Figure 2. Photo used under Creative Commons from diametrik
Figure 3. Photo used under Creative Commons from angus_mac_123
Figure 4. Photo used under Creative Commons from bortescristian
Figure 5. Photo used under Creative Commons from etcname
Figure 6. Image Courtesy SHo Architects
Figure 7. Image Courtesy SHo Architects
Figure 8. Image Courtesy SHo Architects
Figure 9. Photo used under Creative Commons from Andrew Stawarz
Figure 10. Photo used under Creative Commons from kudumomo
Figure 11. Image courtesy SHo Architects

25. Integrated Design: A Computational Approach to the Structural and Architectural Design of Diagrid Structures, Jessica Sundberg Zofchak
Figure 1. Photo used under Creative Commons from Lauren Manning
Figure 2. Photo used under Creative Commons from Lauren Manning
Figure 4. Photo used under Creative Commons from harshil.shah

26. Form Finding: The Engineer's Approach, Fernando Pereira Mosqueira
Figure 1. photo used under Creative Commons from borkur.net
Figure 2. photo used under Creative Commons from Rev Stan
Figure 3. photo used under Creative Commons from priscillajp
Figure 7. photo used under Creative Commons from Tirin
Figure 8. photo used under Creative Commons from yeowatzup
Figure 9. photo used under Creative Commons from Rror

Figure 10. photo used under Creative Commons from Canaan
All other images by author

27. Henderson Waves Bridge, IJP Corporation/ AKT
Figure 1. Image copyright MHJT, courtesy of AKT
Figure 2. Image copyright MHJT, courtesy of AKT
Figure 3. Image copyright MHJT, courtesy of AKT

28. Henderson Waves: A Collaboration, George L. Legendre
All images by author

29. Double Shell, YueYue Wang, Hailong Wu, Zhu Wu
All images by authors

30. Jewel Box, Vera Baranova, Sophia Chang, Bernard Peng
All images by authors

31. Shanghai Expo 2010 UK Pavilion, Heatherwick Studio/ AKT
Figure 1. Image copyright Hufton+Crow, courtesy of AKT
Figure 2. Image copyright Hufton+Crow, courtesy of AKT
Figure 3. Image copyright Hufton+Crow, courtesy of AKT

32. In Collaboration. . ., Katerina Dionysopoulou
Figure 1. Courtesy of Heatherwick Studio
Figure 2. AKT
Figure 3. AKT
Figure 4. AKT
Figure 5. Image courtesy of Heatherwick Studio
Figure 6. Image courtesy of Daniele Mattioli
Figure 7. Image courtesy of Heatherwick Studio

33. Death of the Star Architect, Jennifer Bonner
Image 1. Image courtesy of NASA/JPL-Caltech/CXO/CfA

35. A Necessary Resistance within Architect-Engineer Collaboration, Fai Au
Figure 1. Image courtesy of OMA
Figure 2. Image used under Creative Commons from HeyItsWilliam
Figure 3. Image courtesy of IJP Corporation
Figure 4. Background image courtesy of IJP Corporation
Figure 5a and 5b. Image courtesy of ETH professor Philip Block

36. Interdisciplinary Collaboration: Enabling Architects to Regain Leadership in the Building Industry, Advait Sambhare
All images by author

37. Adelaide Wharf Housing, AHMM/ AKT
Figure 1. Image copyright Valerie Bennett, courtesy of AKT
Figure 2. Image copyright Valerie Bennett, courtesy of AKT
Figure 3. Image copyright Valerie Bennett, courtesy of AKT

38. Collaboration as a Working Process, Morag Tait
Figure 1. Courtesy of AHMM
Figure 2. Courtesy of AHMM
Figure 3. Courtesy of AHMM
Figure 4. Courtesy of AHMM

39. Center for Advanced Architecture, Andrew Pedron, Nathan Shobe, Trisitie Tajima
All images by authors

40. Bridge Gallery, Danxi Zou, Song He, Bo Feng
All images by authors

41. Westfield Student Village, FCB Studios/ AKT
Figure 1. Image copyright Valerie Bennett, courtesy of AKT
Figure 2. Image copyright Valerie Bennett, courtesy of AKT
Figure 3. Image copyright Valerie Bennett, courtesy of AKT

Imprint

Published by
Harvard University Graduate School of Design
www.gsd.harvard.edu
Actar
www.actar.com

Edited by
Hanif Kara and Andreas Georgoulias

Editorial supervision
Melissa Vaughn, Director of Publications

Graphic design
Ulises Chamorro

Production
ActarPro

© of the edition, Actar and Harvard University
Graduate School of Design, 2012
© of the texts, their authors
© of the images, their authors
All rights reserved

ISBN 978-84-15391-08-1
DL B-27040-2012

Printed and bound in the European Union

Distribution

ActarD
Barcelona - New York
www.actar-d.com

Roca i Batlle 2
E-08023 Barcelona
T +34 93 417 49 93
F +34 93 418 67 07
salesbarcelona@actar.com

151 Grand Street, 5th floor
New York, NY 10013, USA
T +1 212 966 2207
F +1 212 966 2214
salesnewyork@actar.com